# Nuclear Weapons and Security: The Effects of Alternative Test Ban Treaties

# Nuclear Weapons and Security

## The Effects of Alternative Test Ban Treaties

EDITED BY

Jonathan Medalia
Paul Zinsmeister
Robert Civiak

Routledge
Taylor & Francis Group

LONDON AND NEW YORK

First published 1991 by Westview Press, Inc.

Published 2018 by Routledge
52 Vanderbilt Avenue, New York, NY 10017
2 Park Square, Milton Park, Abingdon, Oxon OX14 4RN

*Routledge is an imprint of the Taylor & Francis Group, an informa business*

Library of Congress Cataloging-in-Publication Data
Nuclear weapons and security: the effects of alternative test ban
    treaties / edited by Jonathan Medalia, Paul Zinsmeister, Robert
    Civiak.
        p.  cm.
    ISBN 0-8133-8261-0
    1. Nuclear weapons—Testing.  I. Medalia, Jonathan E.
II. Zinsmeister, Paul.   III. Civiak, Robert.
U264.N818   1991
355.8′25119′0287—dc20                                    90-27035
                                                              CIP

ISBN 13: 978-0-367-01643-2 (hbk)
ISBN 13: 978-0-367-16630-4 (pbk)

# Contents

# Tables and Figures

# Acknowledgements

The principal authors of this study were David Cheney and Robert Civiak of the Science Policy Research Division, Warren H. Donnelly of the Senior Specialists Section, and Jonathan Medalia, Charlotte P. Preece, Robert G. Sutter, and Paul Zinsmeister of the Foreign Affairs and National Defense Division. The study was coordinated by Paul Zinsmeister, along with David Cheney, Robert Civiak, and Jonathan Medalia.

We would like to thank many individuals for providing information or commenting on drafts: Ralph Alewine III, Defense Advanced Research Projects Agency; Robert Barker, Assistant to the Secretary of Defense for Atomic Energy; Nicholas Carrera, Arms Control and Disarmament Agency; Cosmo DiMaggio III, Analyst in Science and Technology, Science Policy Research Division, Congressional Research Service; Steven Fetter, John F. Kennedy School of Government, Harvard University; Richard Garwin, IBM Thomas J. Watson Research Laboratory; James Hannon, Warren Heckrotte, and Ray Kidder, all of Lawrence Livermore National Laboratory; Cyrus "Skip" Knowles, R&D Associates; Gerald Marsh, Argonne National Laboratory; Robert S. Norris, Natural Resources Defense Council; James Powell, Sandia National Laboratory; Paul Richards, Lamont-Doherty Geological Observatory; Josephine Stein; Theodore Taylor; and Gregory van der Vink, Office of Technology Assessment.

We would like to extend special thanks to John Immele, Deputy Associate Director for Nuclear Design, and Paul Brown, Assistant Associate Director for Arms Control, both of Lawrence Livermore National Laboratory. They were of great help throughout the study, meeting with project staff, offering detailed comments on drafts, providing documents, answering questions, suggesting others to contact, and in general teaching us about nuclear weapons, nuclear testing, and the relationship between them. They were knowledgeable, helpful, and accessible. They improved this study tremendously.

We are also grateful for the insights and direction given us by Stanley Heginbotham, former Chief of the Foreign Affairs and

National Defense Division, especially for helping us through the "rough spots."

We would like to thank Erin Day, Elena Pappas, and Dianne Rennack for assistance in preparing the manuscript.

Finally, we appreciate the cooperation and patience of Edward Barnes, Miles Gray, Donald Little, and others at the U.S. Department of Energy who conducted a security review of our study.

While many people aided the study effort, the responsibility for any errors of fact, interpretation, or omission rests completely with the authors and the Congressional Research Service.

*J. M.*
*P. Z.*
*R. C.*

# About the Contributors

**David Cheney** is a Senior Associate, Council on Competitiveness, Washington, D.C., and was an Analyst in Science and Technology, Science Policy Research Division, Congressional Research Service, Library of Congress.

**Robert Civiak** is a Budget Examiner, Nuclear Energy Branch, Office of Management and Budget, and was a Specialist in Energy Technology, Science Policy Research Division, Congressional Research Service, Library of Congress.

**Warren H. Donnelly** is a Senior Specialist in Energy Policy, Senior Specialists Section, Congressional Research Service, Library of Congress.

**Jonathan Medalia** is a Specialist in National Defense, Foreign Affairs and National Defense Division, Congressional Research Service, Library of Congress.

**Charlotte P. Preece** is Assistant Chief and Division Specialist in International Security Policy, Foreign Affairs and National Defense Division, Congressional Research Service, Library of Congress.

**Robert G. Sutter** is a Senior Specialist, Foreign Affairs and National Defense Division, Congressional Research Service, Library of Congress.

**Paul Zinsmeister** is a Specialist in National Defense, Foreign Affairs and National Defense Division, Congressional Research Service, Library of Congress.

# 1

# Introduction

*David Cheney, Robert Civiak,*
*Jonathan Medalia, and Paul Zinsmeister*

Efforts among the nations of the world to achieve a total ban on nuclear test explosions have waxed and waned for more than three decades. Although a complete ban has not been achieved, diplomatic negotiations have produced four treaties limiting nuclear explosions: (1) the Limited Test Ban Treaty of 1963, signed by over 100 nations including the United States and the Soviet Union, in which nuclear testing in the atmosphere, in space, and under water is banned; (2) the 1968 Treaty on Non-Proliferation of Nuclear Weapons (NPT), in which some 130 nations have pledged not to acquire nuclear weapons or other nuclear explosive devices; and (3) the Threshold Test Ban Treaty (TTBT) of 1974 and (4) the Peaceful Nuclear Explosions Treaty (PNET) of 1976, in which the United States and the Soviet Union agreed to ban military and peaceful nuclear explosions with an explosive yield greater than 150 kilotons (kt). The latter two treaties have not been ratified, but each nation has undertaken to abide by the 150-kt limit as long as the other does.

For the past few years, the Soviet Union has been calling for negotiations with the United States toward a comprehensive test ban, but has until recently refused to discuss stricter verification procedures for the TTBT and PNET, which was the Reagan Administration precondition for such talks. In another venue, the House repeatedly passed legislation to restrict U.S. weapons tests to below one kiloton so long as the Soviet Union abided by the same restriction. Congressional pressure on the Reagan Administration and progress on bilateral negotiations to reduce the size of U.S. and Soviet nuclear arsenals appear to have spurred both nations to agree to talks with the ultimate objective of the complete cessation of nuclear testing as part of a wide-ranging arms reduction process. The first goal in the talks, which began on November 9, 1987, is agreement on effective verification measures to make it possible to ratify the TTBT and the PNET.

Debate on the test ban issue often has been shrouded in contradictory arguments and emotional appeals. To help remedy this situation, Representatives William S. Broomfield and Beverly Byron asked the Congressional Research Service to "undertake a major study of all aspects of the nuclear testing issue in order to create an objective, factual and analytical basis for Congressional consideration of various nuclear test ban and testing moratoria initiatives." This report responds to that request. It concentrates on three possible kinds of treaties between the United States and the Soviet Union: a comprehensive test ban treaty (CTBT), a low-yield threshold treaty (LYTT) either allowing nuclear test explosions up to 1 kt or allowing tests up to 20 kt (compared to the current 150-kt limit), and a Quota Treaty that would limit the number of nuclear test explosions to some specified number (presumably one or two a year) over a given period, perhaps in conjunction with a threshold below 150 kt.

This report examines several aspects of these treaties. By way of background, it discusses the history of test ban efforts (chapter 2) and nuclear warhead technology, stockpiles, and testing (chapter 3). Next, it examines consequences of these treaties in three key areas: development of new nuclear weapons (chapter 4), maintenance of confidence in stockpiled nuclear warheads (chapter 5), and determination of the effects of nuclear explosions on military equipment (chapter 6). It then considers U.S. ability to verify compliance with various treaties limiting nuclear testing. It concludes by discussing the views of the non-superpower nuclear weapon states and the NATO allies on these treaties and the implications of these treaties for efforts to slow the worldwide spread of nuclear weapons. This report is written from the U.S. perspective because, glasnost notwithstanding, much less is known about Soviet nuclear weapons capabilities and programs.

The balance of this introductory chapter summarizes our findings on the topics in chapters 4-8.

## Implications of a More Restrictive Nuclear Test Ban on Nuclear Weapons

Nuclear testing is an integral part of (1) developing new nuclear weapons, (2) maintaining confidence in existing nuclear weapons, and (3) assessing the vulnerability of military systems to the effects of nuclear weapons. To the extent a test ban treaty is more restrictive than the current regime, it will limit the capability of the United States and the Soviet Union to increase or maintain the military value of their nuclear weapons. Our findings regarding the extent to which different treaties might affect the value of nuclear weapons follow.

- Neither a CTBT, an LYTT, or a Quota Treaty would by itself prevent deployment of nuclear weapons, but each would restrict to varying degrees, the development of new nuclear warheads.

A new test ban alone would not prevent the United States or the Soviet Union from deploying more nuclear weapons of existing types; nor is it likely to prevent deployment of new weapons currently in advanced stages of development. Both countries have nuclear weapons under development. The Soviet Union is flight testing a new ICBM and will probably begin flight testing a new SLBM soon, but there is no unclassified information regarding new warheads for these or other Soviet weapons under development. In the United States, warheads for two weapons are in the final stage of development: the W88 warhead for the Trident II and the W82 warhead for an upgraded 155 mm nuclear artillery projectile. Much of the testing has been done on these warheads. If a test ban treaty precludes further testing, the armed forces likely would have less confidence in the warheads than they desire, but it also is likely that they would have sufficient confidence to deploy either weapon. The W87, planned for use on Midgetman, is fully tested and used on the MX

(Peacekeeper). No further tests are planned with respect to its modification and use on Midgetman if production of the W87 for Midgetman follows production of the W87 for MX with little delay.

Except for those in advanced stages of development, a CTBT would preclude development of new warheads optimized to new delivery systems or missions. Specifically, under a CTBT the United States could not optimize a strategic earth penetrator warhead or a warhead for a maneuvering reentry vehicle (MaRV). Both warheads are currently undergoing feasibility studies. The United States and Soviet Union could, however, unless forbidden by other arms control measures, develop improved weapon systems using existing or slightly modified warheads. For example, the United States could deploy MaRVs or earth penetrator weapon systems using existing or modified warheads, but the resulting systems would be of less military value than would be the case if new warheads optimized to the missions were developed.

The effect of an LYTT on warhead development would depend on the threshold. It is difficult to develop a modern warhead without testing it at a substantial fraction of the intended yield. A 1-kt LYTT would have almost the same effect as a CTBT on the development of current generation strategic weapons,[1] but it would permit unhindered development of some tactical weapons. The so-called neutron bomb, for example, has a yield in the 1-kt range. A 20-kt LYTT could allow considerable development of strategic warheads. For example under a 20-kt LYTT, it might be possible to develop warheads for earth penetrators and MaRVs that could have significant strategic value if sufficient accuracy of delivery can be achieved to compensate for limited warhead yield.

The effect of a Quota Treaty would depend on the number of tests permitted and whether it includes a threshold below 150 kt. A quota of one or two tests a year would be similar to a CTBT with respect to developing new warheads to the extent that other testing requirements (e.g., to maintain confidence in existing warheads or to protect military hardware against nuclear effects) were given priority.

- A CTBT or Quota Treaty would for all practical purposes prevent the development and deployment of future generations of nuclear weapons. Some such weapons might be developed under a 20-kt LYTT and to a much more limited extent under a 1-kt LYTT.

For this report, future generation nuclear weapons are those that enhance, suppress, or direct specific outputs from nuclear explosions, as well as nuclear weapons concepts yet unknown. A virtually unlimited variety of such future weapons is theoretically possible, but few might have military uses.

A CTBT would for all practical purposes prevent development of future generation nuclear weapons because extensive nuclear testing would be required to understand and make use of their properties. A Quota Treaty allowing one or two tests per year would have a similar result, although it is conceivable that the United States or Soviet Union might be able to develop a weapon over decades if it were given priority over other testing requirements.

A 20-kt LYTT, and to a much lesser extent a 1-kt LYTT, would allow research on future generation weapons, but whether effective weapons could be developed or deployed is problematical. Theoretically, low yield weapons could be used effectively in space or in the upper atmosphere, where damaging radiation or particles can travel large distances. Development of such weapons is also constrained by the Limited Test Ban Treaty, which prohibits nuclear testing in the atmosphere or space.

Two future generation weapons concepts which have had underground tests are the hypervelocity pellet gun (nuclear shotgun) and the X-ray laser. Both weapons would be used to destroy targets in space. The hypervelocity pellet gun would shoot small particles at extremely high velocities toward a target. It appears that a 1-kt LYTT would permit development of this weapon, since one kiloton is more than enough energy for this task. An X-ray laser would absorb X-rays from a nuclear explosion and re-emit them in narrow beams toward multiple targets. The efficiency of this process is currently very low and even with substantial gains it appears that high yields (perhaps even above the current 150-kt testing limit) may be needed for an effective weapon. Research on many aspects of an X-ray laser could proceed under a 20-kt LYTT and on some aspects under a 1-kt LYTT.

- A CTBT (and possibly a 1-kt LYTT) would reduce U.S. and Soviet capability to test for the effects of X-rays and the synergistic effects of all products of a nuclear explosion on military systems. A Quota Treaty and a 20-kt LYTT could permit such tests. Other nuclear effects can be simulated adequately by nonnuclear means.

Effects tests are used to understand the effects of nuclear weapons and improve the ability of military systems to survive nuclear explosions. The more a treaty restricts using nuclear explosions for effects testing, the more it serves to constrain development of new weapons generally.

Most nuclear effects can be simulated adequately by nonnuclear means. The major result of limiting the use of nuclear explosions for effects tests would be to diminish U.S. and Soviet capability to assess and improve the resistance of equipment such as military satellites, reentry vehicles and potential SDI systems to nuclear generated X-rays and to the synergistic effects of simultaneous multiple nuclear effects.

The United States conducts one or two nuclear effects tests a year. To sustain an equivalent capability under a CTBT would require a significant improvement in producing X-rays by nonnuclear means. Pending these improvements, extra measures could be taken to protect systems against X-rays (e.g., extra shielding, more resistant electronics), although such measures may impose performance penalties.

Effects tests could be conducted under a Quota Treaty (although they would have to compete with other test requirements) and under a 20-kt LYTT. In principle, most of the knowledge of the effects of nuclear explosions on military equipment gained from underground tests could be obtained from explosions of devices with yields of less than 1 kt. Whether such nuclear devices could be developed under a 1-kt LYTT is not evident from open sources.

A halt in effects testing is not likely to affect deployment of the Trident II (D-5) and Midgetman missiles because effects tests will have been completed on their reentry vehicles, which are the stage most vulnerable to nuclear attack. Other tests are scheduled to assess the vulnerability of the D-5 and Midgetman missiles to Soviet nuclear-armed ballistic missile defenses. Precluding these tests could reduce confidence in the ability of these systems to survive the effects of such warheads, but D-5 and Midgetman will face only a severely constrained defense (100 interceptors around Moscow) as long as the Antiballistic Missile Treaty holds.

- Any test ban more restrictive than the current regime would over time reduce confidence among U.S. and Soviet leaders that some warheads in their respective stockpiles would perform as designed. This effect would become more pronounced over many years.

The performance of stockpiled warheads can degrade or become suspect over time as materials deteriorate with age or as design or production defects are discovered. To assure they will be ready on demand, stockpiled warheads are monitored through an extensive program of disassembly, inspection, and nonnuclear testing. These procedures usually suffice to detect and fix problems. Occasionally, nuclear tests on warheads under development reveal problems with warheads in the stockpile. On rare occasions, tests have been used to look for flaws in deployed warheads; in at least one case, a problem was found. Sometimes, tests have been used to evaluate problems or to verify performance after a fix.

Any test ban more restrictive than the current regime would, over time, lead to some reduced confidence in the reliability of warheads because it would preclude some confidence tests to address specific warhead problems and because the full range of testing helps warhead designers maintain their expertise to identify, evaluate, and fix warhead problems.

It appears likely, however, that a substantial loss of confidence in warhead reliability would not occur for many years. The current U.S. inventory contains 26 distinct warhead types.

For 12 of the 26 warheads, nuclear tests subsequent to their deployment identified or confirmed a problem or demonstrated that a warhead would work after a problem had been corrected. For all but two of these warheads, the test or tests were done within four years of their deployment. Following that initial shakeout period, warheads have remained in the stockpile for 20 years or more without problems for which nuclear tests were conducted, so that the reliability of tested nuclear warheads appears sustainable for long periods without further nuclear testing.

- Confidence in the present U.S. stockpile of nuclear warheads is high. Several measures could slow the loss of stockpile confidence under a test ban. These measures, however, raise confidence issues of their own.

At present, a warhead is much less likely to malfunction than other weapon system components such as the rocket engine or the guidance system. Under a more restrictive test ban, some problems in aging warheads could be corrected by replacing deteriorated or suspect warheads or their components with warheads or components newly remanufactured to original specifications. This approach would avoid some uncertainties resulting from an inability to test. Other uncertainties, however, would arise: (1) Flaws in the original design or rapidly occurring degradation of components or materials could not be fixed by remanufacturing. (2) It would be difficult to have full confidence in remanufactured warheads without testing them because slight variations in materials, components, and the production process could cause the remanufactured warheads to fail.

Difficulties of remanufacture could be eased by building more robust warheads than existing ones and testing them before a more restrictive test ban took effect. For example, warheads containing more nuclear material or high explosive would generally be less sensitive to deterioration. This approach would not, however, guard against some types of deterioration or against many design and production defects. Moreover, a more robust warhead generally equates to a heavier or less efficient warhead, which could necessitate other adjustments such as arming missiles with fewer warheads or building larger missiles.

In some cases, one type of warhead could be used to replace another that had become suspect. Such substitutions are often possible with existing warheads. As examples, bombers carry five types of strategic nuclear bombs, Minuteman III currently carries two different warhead types, and Trident II may 2carry two warhead types. If confidence in a warhead type declines, a substitute could be manufactured and used. Substitution, however, entails two limitations. First, national capability is often reduced by substitution. The substitute warhead may be less capable by itself or in its new role; alternatively, it may have more capability than is needed, which wastes nuclear material, raises cost, or reduces other aspects of capability. (For example, replacing a lower-yield warhead with a heavier, typically higher-yield, warhead would reduce missile range or reduce the number of warheads a missile could carry.) Second, manufacturing new warheads without testing them is subject to the uncertainties noted above.

Several treaties would result in less loss of confidence in the reliability of stockpiled warheads than would a CTBT. In theory, a Quota Treaty permitting one or two tests per year could be sufficient for evaluating suspected problems in stockpiled warheads and for testing modifications that need to be made, given that one or two tests a year are conducted for these purposes now. Under a quota, however, the collective expertise presently embodied in the weapons laboratories would decline over time, which eventually could reduce confidence in the reliability of the stockpile. A 20-kt LYTT could allow a vigorous testing program, including full-yield testing of the primaries of thermonuclear warheads, the stage in which most reliability problems occur. Such testing could be sufficient to maintain expertise for designing primaries and to evaluate and fix most problems, except for the rare instance of a problem in the secondary of a thermonuclear weapon. A 1-kt LYTT would permit some testing on matters affecting stockpile confidence, and would do more to maintain skills of weapon designers and retain them at the weapon laboratories than would a CTBT.

- Over the long term, a more restrictive test ban would cause U.S. and Soviet national leadership to place less value on the capability of nuclear weapons to perform counterforce missions.

Counterforce nuclear weapons can be used to destroy an enemy's strategic nuclear forces, including their command structure, in a first strike or preemptive strike. The attacker would hope that the destruction would be so thorough that the enemy would be unable or unwilling to launch his remaining forces in retaliation. The national leadership must have high confidence in their counterforce weapons to attempt such a mission. A more restrictive test ban would prevent or severely impede each side from developing improved counterforce weapons to further threaten the other side's nuclear forces and would cause some loss of confidence in the reliability of stockpiled warheads over time. As a result, each side's leadership would have less confidence in its ability to gain a decisive military advantage from a first strike, making such an attack less likely. (Of course, it is not possible to gauge with any precision what the confidence level of each side for a first strike is now.)

On the other hand, a more restrictive test ban would have less effect on each side's confidence in the capability of its nuclear weapons to devastate the other side's cities and industries. Nuclear weapons need not be as reliable for this "countervalue" mission as they must be for the counterforce mission. Deterrence would rest on the perception by each side that enough of the other side's nuclear weapons would work to make an attack too risky.

- A more restrictive nuclear test ban, including a CTBT, would not by itself stop the United States or the Soviet Union from taking steps that could increase the threat against the other's nuclear weapons or that could make their own weapons more secure. These actions could alter the strategic balance between the two nations.

A number of measures could be taken by either side under a nuclear test ban that would threaten the other side's nuclear weapons, such as improving the accuracy of missiles aimed at hardened silos, improving technologies for finding and destroying strategic submarines, and deploying a nonnuclear ballistic missile defense (BMD) of the type undergoing research under the Strategic Defense Initiative (SDI). Conversely, nuclear weapons might be made more secure by expanding mobile basing of missiles, further hardening of missile silos, operating strategic submarines more quietly and over larger ocean areas, or by developing and deploying a BMD to protect ICBM sites.

Restricting the development of new warheads through a new test ban treaty would not by itself stop the strategic competition between the United States and the Soviet Union. It is impossible to know where this competition might lead in the future, but with or without a new test ban treaty, nonnuclear measures could have a large effect in determining that outcome. On the other hand, a nuclear test ban treaty might provide impetus for further arms control measures that could affect the strategic balance and strategic stability. Whether that result is perceived as good or bad depends upon one's perceptions of how U.S. strategic interests are best protected as discussed below.

## Potential for Adequate Verification
## of a Test Ban Treaty

Disagreement between the United States and the Soviet Union on the means of verifying mutual compliance has long been a barrier to a further test ban treaty. To verify compliance with any such treaty, the United States must be able to detect and identify nuclear explosive tests not permitted by the treaty.

Two developments have combined to make this thorny issue more susceptible to resolution today. First, monitoring technologies have improved. Clandestine underground explosions are generally considered the most difficult to monitor, but progress in seismic methods has greatly improved the ability to detect and identify distant underground nuclear explosions. Second, the Soviet Union has shown a new willingness to accept highly intrusive monitoring procedures. The ability to detect cheating cannot be perfect, however, and there will always be some uncertainty in verifying treaty compliance.

Accordingly, the political debate asks, "How much verification is enough?" There are two standards set for a verification regime. One, followed by the Nixon, Ford, and Carter Administrations, is that of "adequate verification," defined as the ability to detect cheating of significance to the strategic balance in time to take appropriate action. The other, "effective verification," was used by the Reagan Administration and generally required the detection of a single significant violation. The choice between these standards rests on differing political judgments, such as how much we can trust the Soviet Union, and technical judgments, such as the extent to which a single clandestine nuclear test could alter the strategic balance. The choice of standard goes to the heart of the political debate, however, because it strongly influences whether the United States and Soviet Union can conclude a treaty that limits testing more than the current treaty regime does.

- Seismic monitoring networks are capable of detecting and identifying clandestine nuclear explosions of a few kilotons in support of a CTBT or other treaty. The lowest limit is debated but there is agreement that seismic methods cannot be depended on to reliably identify clandestine tests below 1 kiloton.

Observation of seismic waves is the best method available for monitoring underground explosions within a large area. "Adequate" verification of a treaty banning small nuclear explosions would require both sides to agree to and observe procedures whereby the other side could install, calibrate, maintain, and monitor seismic stations on the other's soil; "effective" verification would require more extensive and more intrusive monitoring procedures. Technical specialists disagree over the lowest yield at which nuclear explosions could be detected reliably over large areas and distinguished from earthquakes. They also disagree over the number of stations that would be required in the Soviet Union to effectively monitor Soviet compliance.

An optimistic view is that 25 simple seismometers within the Soviet Union, complemented by a similar number of stations in neighboring countries, could detect and identify 1-kt nuclear explosions with high confidence, even if the seismic signals generated by the explosion are intentionally reduced by conducting the explosion in soft rock or a large underground cavity (cavity decoupling). A pessimistic view is that a network of 15 seismic arrays (each with several seismic stations) within the Soviet Union could not conclusively identify all cavity decoupled explosions of a few kilotons.

Under a CTBT, the United States could not be certain the Soviets were not conducting decoupled nuclear tests of less than 1 kt. Even with the most sensitive seismic monitoring network, such tests could not be easily distinguished from chemical explosions used for mining and construction. Nonetheless, an extensive seismic network complemented with other monitoring methods (e.g., satellites, human intelligence) introduces a significant element of risk to a would-be cheater. In addition, a series of clandestine tests at explosive yields below seismic detection capability is more likely to be detected than a single test. Whether verification of a CTBT is adequate on these terms depends on how one judges the military value of clandestine tests of less than 1 kt compared to the overall value of the treaty.

- Verification of compliance with a 1-kt LYTT may be possible, but many technical and political issues must be resolved.

To verify compliance with a 1-kt LYTT, it is necessary to accurately determine the yield of explosions at designated test sites and to detect and identify clandestine tests outside designated test sites. The extent to which these tasks can be accomplished has not been fully resolved by the technical community. In addition, political questions such as how much uncertainty in verifying treaty compliance with a 1-kt LYTT and how much intrusion from monitoring programs are acceptable to both sides remain largely unaddressed.

An optimistic view of verifying a 1-kt LYTT holds that (1) with a suitable in-country seismic system, any clandestine tests above 1 kt away from designated test sites would be detected and that (2) with negotiated controls on the locations in which testing is permitted and with well-calibrated seismic methods for estimating yields (achieved, for example, by allowing each country to set off and monitor explosions of its nuclear devices at the other's test site), the uncertainties in yield estimates would be sufficiently low to prevent significant military gains from testing above the threshold.

A more pessimistic view holds that (1) even with extensive in-country seismic stations, it would be impossible to be sure that some seismic signals appearing to be earthquakes or chemical explosions were not really decoupled nuclear explosions of a few kilotons and that (2) there are many unresolved technical difficulties in estimating the yields of small explosions (using either seismic methods or on-site measurement techniques such as the CORRTEX method proposed by the Reagan Administration to verify compliance with the TTBT) that may lead to questions about both Soviet compliance and the adequacy of verification. In this view, there is currently insufficient evidence on which to determine if a verification regime is possible because there is little data on estimating yields of small nuclear explosions in geologic areas similar to the Soviet test sites.

- The technology is available to monitor compliance, using seismic means, with a 20-kt LYTT.

Existing monitoring stations outside the Soviet Union could detect and identify clandestine tests outside of designated test sites well enough to meet the "adequate verification" standard for a 20-kt LYTT. There would be some doubt, however, as to the precision with which these stations could determine yield of explosions near 20 kt. Accordingly, application of the "effective verification" standard could require negotiated treaty provisions for in-country seismic stations, better calibration of yield estimation methods, and improved information about the geology of the Soviet test site. These provisions could reduce uncertainties in estimating yields of explosions near 20 kt to below the level currently associated with 150-kt tests. Employing the CORRTEX method of estimating yields would increase the precision of yield estimates but would be more intrusive than seismic methods alone.

- Verification of a Quota Treaty might require on-site inspection to determine whether several explosions are conducted simultaneously.

A Quota Treaty would likely have an upper threshold and may also have a lower threshold. Verifying a Quota Treaty could require measures discussed above for determining compliance with upper and lower thresholds. In addition, if the quota limits the number of explosions rather than the total yield of any number of explosions, on-site inspection could be required to determine whether more than one explosion is conducted during a single test.

## International Implications of a More Restrictive Nuclear Test Ban

- The long-term success of global efforts to constrain nuclear warhead development requires participation by all nuclear weapons states. At this time, however, the non-superpower nuclear weapons states oppose further restrictions on nuclear testing.

The willingness of the recognized non-superpower nuclear weapons states--France, United Kingdom, People's Republic of China--to participate along with the United States and the Soviet Union in a more restrictive nuclear test ban will be essential to the long-term success of global efforts to constrain nuclear warhead development. At this time, however, these three countries oppose further restrictions on nuclear testing, arguing that their nuclear arsenals are small compared to those of the superpowers and, given the current global nuclear balance, that the superpowers should first make significant cuts in their nuclear and conventional forces before expecting other nuclear weapons states to accept additional testing restrictions.

The governments of France, Britain, and China maintain that as long as nuclear missiles remain the foundation for global deterrence, they must continue to test to develop new systems in order to keep their deterrents credible. Furthermore, the British and French governments, as well as other European allied nations, are concerned that further U.S./Soviet testing restrictions would undermine confidence in America's nuclear guarantee to Western Europe. In other words, these three nuclear weapons states view a test ban as following on significant superpower arms reductions and not as a vehicle to generate other arms reductions agreements.

- A CTBT would support the Treaty on Non-Proliferation of Nuclear Weapons (NPT) and might help dissuade nonnuclear weapons states from acquiring their own nuclear arsenals. A less restrictive test ban is unlikely to achieve those results.

In recent years, many countries grouped under what is referred to as the "non-aligned bloc" have complained that their faithful abstinence from pursuing nuclear weapons as pledged under the NPT has not been matched by corresponding progress of the nuclear weapons NPT states towards disarmament. The longer the superpowers are perceived as stalling on a CTBT, the greater the prospects that some nonnuclear weapons states will impede efforts to extend the NPT as an effective restraint on nuclear weapons proliferation. This same group of nations would generally regard any test ban less restrictive than a CTBT as another instance of the nuclear weapons states trying to preserve their monopoly on nuclear weapons. Therefore, less restrictive agreements are unlikely to help efforts to extend the NPT.

Nations now considered to be at or near the threshold of nuclear weaponry, including Argentina, Brazil, India, Israel, Pakistan, and South Africa, would be constrained by a CTBT only if they joined such an agreement--a distant prospect--or to the extent that international measures aimed at stopping these and other non-weapons states from acquiring nuclear weapons, such as the NPT, are strengthened by a CTBT.

# Notes

1. For this study, the term current generation nuclear weapon means those for which no special effort is made to modify the form or direction of the explosive energy release. Future generation weapons, discussed below, either enhance, suppress, or direct particular effects of nuclear explosions.

# 2

---

# The Test Ban Debate:
# Forty Years of New and
# Recurring Themes

*Jonathan Medalia*

The evolution of the nuclear test ban debate can be seen as a competition between changing sets of arguments, with differing arguments gaining and losing prominence over time. The debate since 1945 can be divided into six periods, each with its own political and military context and with differing concerns expressed by protagonists. The periods, their context, and the dominant concerns regarding nuclear weapons and their testing are: (1) 1945-1950: Cold War/Destructiveness; (2) 1951-1957: Rearmament/Fallout; (3) 1958-1963: Competition/Verification; (4) 1964-1976: Detente/Proliferation, Reduction in Permitted Test Yields; (5) 1977-1980: Arms Control Initiatives/Weapons Reliability; (6) 1981-1987: Competition/New Weapons. This chapter presents an overview of the test ban debate organized around these periods and themes.[1]

## 1945-1950: COLD WAR

### Nuclear Concern--Destructiveness

The debate over nuclear disarmament began shortly after the bombings of Hiroshima and Nagasaki. Public concern at this time focused almost exclusively on the destructiveness of nuclear weapons. Fearing that nuclear weapons could destroy humanity, many favored abolishing all nuclear weapons, not just limiting nuclear testing.

During this period, fear of nuclear weapons led the United States and Soviet Union to attempt to limit them. Not surprisingly, the plans put forth on nuclear weapons reflected each side's fears and strengths. In 1946, at the United Nations, the United States proposed the Baruch plan, which envisioned an international authority to control "all phases of the development and use of atomic energy." Under the plan, manufacture of atomic bombs would stop, existing bombs would be disposed of, and punishments would be imposed for viola-tors.[2] In response, the Soviet Union set forth the Gromyko plan, which would ban the production, storage, and use of atomic weapons, and called for the destruction of all atomic weapons within three months after the agreement entered into force. Within six months, the parties would agree to penalties for violators.[3]

Efforts were also made within the United States to unilaterally eliminate or limit nuclear weaponry. In 1946, prior to the first scheduled U.S. nuclear tests in the Pacific, a Senate resolution proposed that the United States cancel the Pacific tests, and a House resolution proposed that the United States cease manufacture of atomic bombs and cancel the Pacific tests.[4] Prominent U.S. religious and scientific groups endorsed these proposals. When efforts were made in the late 1940s and early 1950s to develop the hydrogen bomb, many scientists recommended a halt to its development before it could be tested. Vannevar Bush, who headed the Office of Scientific Research and Development in World War II, was the first to advance formally the idea of a hydrogen bomb test ban agreement on the grounds that the United States should not take responsibility for introducing such a frightful new weapon.[5]

U.S.-Soviet tensions overpowered these activities. With the onset of the cold war, the first Soviet nuclear explosion in 1949, and the Korean War, initial repugnance to nuclear weapons felt by many in the United States gave ground to a perceived need to build more and better nuclear weapons. In 1950, President

Truman placed development of thermonuclear weapons on a crash basis. The speed with which the Soviet Union was able to detonate its first nuclear weapon also sparked concern that other nations would obtain these weapons.

The early debate over whether the destructiveness of nuclear war is best averted by abolishing nuclear weapons, limiting them, or by building new ones to deter their use has lasted throughout the controversy over nuclear testing.

## 1951-1957: REARMAMENT

### Nuclear Concern--Fallout

This period was characterized by U.S. and Soviet rearmament and continued mistrust. The Korean War, the Soviet post-Stalin military buildup, the crushing of the Hungarian uprising, the Suez crisis, Sputnik, and the introduction of ICBMs promoted bellicosity between the superpowers. The two nations used the chief arms control negotiating forum, the United Nations Subcommittee on Disarmament, mainly for propaganda while discussing the elusive concept of "general and complete disarmament."

During this period, the U.S. nuclear test program grew dramatically, from 8 announced explosions from 1945 through 1948 and 0 in 1949 and 1950, to 16 in 1951, 10 in 1952, and 11 in 1953.[6] Nuclear weapon technology proceeded rapidly, with deployment in the early 1950s of many types of tactical nuclear weapons, such as nuclear artillery shells, naval depth bombs, atomic demolition munitions, and battlefield missile warheads, and deployment of the first thermonuclear bomb in 1953.[7] The U.S. stockpile of nuclear weapons grew dramatically during the early 1950s.

In 1954, the United States conducted the "Castle" series of tests of thermonuclear explosives in the Pacific. The "Bravo" test of March 1, 1954, the first U.S. test of a thermonuclear weapon,[8] was the first to demonstrate the lethal effects of fallout at long ranges. It spread fallout in lethal concentrations over more than 7,000 square miles,[9] exposing to radiation 28 U.S. servicemen stationed on an island 150 miles from the explosion, the crews of a Japanese fishing boat and two U.S. Navy ships, and the inhabitants of an island 100 miles from the explosion.[10]

The Bravo test brought fallout to public attention, and spurred a worldwide wave of protest. During 1954 and 1955, India's Prime Minister Nehru proposed a "standstill agreement" to halt all nuclear testing, a group of Asian Prime Ministers issued a call to stop testing and negotiate a total ban under U.N. auspices, and the Bandung Conference of Asian and African States demanded a moratorium on nuclear tests. Albert Einstein, Albert Schweitzer, and Pope Pius XII called for an end to nuclear tests.[11] Domestically, the Chairman of the Atomic Energy Commission suggested a ban on very large tests, and the Federation of American Scientists proposed banning all thermonuclear tests. Adlai Stevenson, the 1956 Democratic presidential candidate, also called for cessation of thermonuclear tests.[12] Widely publicized hearings before the Senate Foreign Relations Committee brought out the dangers to the food cycle of strontium-90, the expected increase in incidence of handicapped babies and leukemia victims, and other consequences of radioactive contamination of the atmosphere.[13] The dangers of fallout led many people to want a test ban as a way to end fallout.

During the mid-1950s, world opinion urged the United States and Soviet Union to cease testing. At the same time, developments in nuclear warhead technology, the prospects for such technical advances as intercontinental ballistic missiles and ballistic missile submarines, and the continuing cold war exerted pressure to continue testing.

On May 10, 1955, the Soviet Union made a major change in its disarmament proposals. It proposed three courses of action: political settlement of the cold war; general and complete disarmament; and partial measures of arms control, in explicit recognition of the impossibility of verifying the destruction of all nuclear weapons.[14] While the second and third courses were at odds except in the long run, the Soviet Union indicated a willingness to proceed along either.[15] As one partial measure, it included for the first time in its proposals a ban on nuclear

testing under international supervision.[16]   This proposed ban
was in part a response to public concern over fallout.[17]

The Western powers praised the new Soviet flexibility, but
were slow in responding.  The summit meeting of July 1955,
with the heads of France, the United Kingdom, the United
States, and the Soviet Union, focused on the central importance
of an inspection system for a disarmament plan, though the
western powers and the Soviet Union disagreed on the system.[18]
In September 1955, at a meeting of the U.N. Disarmament
Commission, Harold Stassen, the U.S. deputy representative,
further stressed the importance of an inspection system for both
the general disarmament issue and for one of its components,
the testing issue.[19]

The difficulties for the United States were in deciding on
whether to pursue general disarmament or narrower measures;
which narrower measures to pursue; the link between nuclear
testing, a narrower measure, and inspection for disarmament;
and the desire for nuclear test limits to halt fallout vs. the
desire to test to develop new weapons.[20]  Moreover, the U.S.
mechanism for formulating policy on arms control and
disarmament was poorly organized.  No one entity had been
empowered to develop and negotiate the U.S. position.
Recognizing this, President Eisenhower initiated a policy review
and appointed Harold Stassen as his Special Assistant for
Disarmament.  A subsequent, more formal response was the
creation of the Arms Control and Disarmament Agency in 1961,
with even the name indicating the persistence of tension
between general disarmament and partial measures.

Given the need for trust between the West and the Soviet
Union to achieve general disarmament, the absence of such
trust, and the impossibility of assuring that all nuclear
warheads on either side had been dismantled, it is not
surprising that nothing came of the general disarmament
schemes of the late 1940s and early 1950s.  Nonetheless, the
two sides were unwilling to abandon attempts to negotiate
agreements limiting forces because of the propaganda
disadvantage that would obtain by so doing, the potential
security value of some limits, and the value that continued
negotiations could have for reducing East-West tensions.

Narrower measures were one arena in which negotiations might achieve valuable results. In particular, a test ban would stop fallout.

Accordingly, effort shifted to narrower measures beginning around 1955. One sign of the shift was that in 1957, the Subcommittee of the U.N. Disarmament Commission, composed of Canada, France, the United Kingdom, the United States, and the Soviet Union, paid particular attention to a nuclear test ban. The western powers, which previously had linked a test ban to a broader disarmament agreement, "hinted that they might accept a loosening of the tie between the test ban issue and other measures of disarmament."[21]

## 1958-1963: COMPETITION

### Nuclear Concern--Verification

Superpower relations became more stable in the late 1950s, marked by Khrushchev's "peaceful coexistence" theme of 1956-1959.[22] The U-2 incident of 1960, however, shattered that stability and reactivated hostility that culminated in the Cuban missile crisis of 1962.

Fallout remained the major domestic concern regarding nuclear testing, which made a test ban a salient political issue. In 1958, for example, there was considerable debate over the biological consequences of fallout.[23] Edward Teller, the "father of the hydrogen bomb," denied that fallout posed a peril. Linus Pauling, a Nobel laureate in biology, said that radioactive carbon from previous tests could cause defects in five million children and millions of cancer cases over the next 300 generations. Public concern was heightened by an Atomic Energy Commission report that New York City had the highest level of strontium-90 in the world, by another report that nuclear weapon tests caused a one-third increase of strontium-90 in people, and by a report by the U.N. Scientific Commission on Radiation Effects that fallout posed a genetic peril. Suggestions that people eat more cabbage and broccoli to increase resistance to fallout, or that adding calcium to farmlands would reduce crop intake of strontium-90, hardly allayed public concern. Concern over the known effects of radiation was no doubt intensified by fear of the unknown (large-scale fallout had first occurred only a few years earlier) and by the conflicting claims regarding consequences of radiation.

Both sides exploited this concern by offering various proposals for test bans, including means of monitoring compliance. Some, as noted, were set forth in 1955-1957. On March 31, 1958, the Soviet Union began a moratorium on nuclear testing. It proposed that other nuclear powers do the same, and warned that it would feel free to resume testing if they continued testing.[24] While the Soviet Union made this proposal just after it had concluded a series of nuclear tests and just before the United States was about to start one, it obtained a public relations coup.

The proposal put the United States in a difficult position because of bad timing and, especially, lack of a test-ban policy. The United States had not decided whether to pursue a test ban alone or as part of a broader agreement. It required that a test ban be adequately verifiable, but had not defined "adequate." The government had no organization for coordinating test-ban policy.

On April 28, in a letter to Khrushchev, Eisenhower reiterated a proposal to convene a panel of technical experts to address verification. The proposal offered the propaganda benefit of moving toward test-ban negotiations, and it temporarily finessed the muddled state of U.S. test-ban policy. (Secretary of State Dulles felt that the Soviet Union was unlikely to accept the proposal.)[25] At the same time, the proposal had value beyond propaganda. An agreement would require means of monitoring compliance. At issue were details of the monitoring system and the level of confidence in monitoring capability needed. The United States required something more than the uninspected moratorium the Soviet Union proposed; the conference might help define what was needed.

Eisenhower proposed that a panel of experts from East and West address the verification issue. He noted: "Studies of this kind are the necessary preliminaries to putting political decisions actually into effect. ... with the practicalities already worked out, the political agreement could begin to operate very shortly after it was signed and ratified."[26]   Thus, when the Soviet Union accepted Eisenhower's proposal, on May 9, the United States was in effect committed to the conference and to a negotiation based on its conclusions.

The United States viewed the Conference of Experts as primarily technical.   Secretary Dulles said that, regarding guidance to the U.S. experts,

> They are to come to their own conclusions as to what is necessary to detect an explosion .... we have given them complete authority to work on this matter as a purely scientific technical matter ... I do not anticipate that there will be any need for political guidance.[27]

The conference opened July 1, 1958, at Geneva.   It comprised two panels:  from the East, scientists from the Soviet Union, Czechoslovakia, Poland, and Rumania; from the West, scientists from the United States, Canada, France, and the United Kingdom.  The experts tried to arrive at a system for monitoring nuclear tests in various environments.  The system was to reflect technical requirements, not political considerations such as intrusiveness.  Nonetheless, the negotiations demonstrated the political character of verification arrangements.  The United States proposed a system with 650 control posts (stations for monitoring a test ban) worldwide; the Soviet Union, one with 100 to 110.  The East held that ground control posts sufficed for collection of radioactive debris; the West countered that aircraft were needed to improve the chance of gathering samples.  The West stressed the difficulties of monitoring a test ban; the East emphasized that technical progress would ease those problems over time.  The final communique noted:

> The Conference reached the conclusion that it is technically feasible to set up, with certain capabilities and limitations, a workable and effective control system for the detections of violations of a possible agreement on the world-wide cessation of nuclear weapons tests.[28]

The technical findings of the conference led President Eisenhower to propose that negotiations on a comprehensive test ban begin October 31, 1958. The Soviet Union agreed. Nonetheless, the United States and United Kingdom continued to test. In response, the Soviet Union resumed testing in September 1958. The United States and United Kingdom began a one-year moratorium, subsequently extended, on nuclear testing on October 31; the Soviet Union tested for a few days thereafter, then began its own moratorium. The moratoria held until September 1961.

The Conference on the Discontinuance of Nuclear Weapon Tests was held at Geneva between the United States, the United Kingdom, and the Soviet Union from October 1958 to January 1962. At the conference, verification was the main issue, and underground nuclear tests posed the main problem for verification. Tests in the atmosphere and under water were much easier to monitor. The conference did not focus on space tests, even though they were harder to detect than atmospheric and underwater tests, perhaps because previous negotiations had focused on underground tests and the first satellite had been launched only in 1957. Given the poor capability of seismology at the time to detect and identify low-yield underground tests, the limited experience with underground nuclear testing, and the dearth of photoreconnaissance intelligence on the Soviet Union, the United States sought treaty provisions to monitor limits on underground tests with on-site inspections, in which inspectors would search the site of a suspected nuclear test for evidence of testing.

Two verification issues proved the main stumbling blocks to a treaty. One was the organization for monitoring compliance (the "control" organization). In essence, the Soviet Union wanted to require the concurrence of the inspected nation on all activities of the organization, while the West wanted the organization to operate without such restraint.

On-site inspections were the second stumbling block. Negotiators discussed such details of inspections as who would control the inspectors, how large the team could be, the nationalities of inspectors, how long the inspectors could stay, how large an area they could inspect, and who would provide the aircraft for transporting them. A key issue was the number of inspections permitted each year. The United States, fearing Soviet cheating, wanted many inspections; the Soviet Union, fearing that the United States wanted inspections for espionage, preferred none but was willing to accept a few inspections on its territory and wanted to exercise control over them.

The verification system set forth in the Conference of Experts report could detect most underground tests of seismic magnitude 4.75, thought to be the signal produced by a 5-kt explosion. Thus the United States was in effect willing, for reasons noted on page 26, to negotiate a treaty in which there would be little ability to detect underground tests below that yield. There were thought to be 20 to 100 earthquakes a year in the Soviet Union that could not be confidently distinguished from 5-kt underground explosions on the basis of seismic data. The Conference of Experts report of August 1958 raised the possibility of inspecting all of them.[29]

On the basis of underground tests done in October 1958, just before its moratorium, the United States concluded by January 1959 that explosions of a given yield produced seismic signals of lower magnitude than previously thought and that there were about twice as many earthquakes as estimated earlier, increasing the difficulty of identifying underground explosions.[30] The United States was unwilling to accept a CTBT, fearing that the Soviet Union could test below 20 kt while the United States would not. To avoid this risk, the United States proposed early in 1960 a treaty permitting lower-yield underground testing: a threshold treaty banning underground explosions of magnitude greater than 4.75 and banning all atmospheric and underwater tests. The proposal envisioned about 20 inspections a year.[31] The Soviet Union accepted this idea if there were a moratorium on smaller tests. The United States, in turn, accepted. The two sides, however, proved unable to agree on the term of the moratorium.

They also could not agree on the number of inspections. Responding to the U.S. call for about 20 a year, the Soviet Union in July 1960 proposed two or three on-site inspections annually. The U.S. position in February 1963 was that at least seven inspections per year were needed.[32] It was ostensibly this gap between three and seven--with neither President Kennedy nor Premier Khrushchev feeling in command of the political support needed to close it--that prevented a CTBT.

The United States saw inspections as a deterrent to cheating because they posed for the Soviet Union the risk that a clandestine test might be discovered. Nonetheless, on-site inspections, despite the focus on them, would have been of little technical value for detecting violations. Hearings before the Joint Committee on Atomic Energy in April 1960 brought out the problems facing inspections.[33] The status of seismology was such that one witness testified, "I could not at present say what [seismic] features are reliable indications of earthquakes. I do not think anybody can."[34] Another witness testified that even with thousands of seismometers on Soviet territory, there could be thousands of seismic events (earthquakes or explosions) in the Soviet Union each year that could not be confidently identified as one or the other on the basis of seismic data.[35] Thus, thousands of on-site inspections on Soviet territory might have been needed each year. These inspections would have encountered great technical difficulties: the need to search up to 200 square miles, the need for large inspection teams, and the possible need to spend years drilling holes in a small area to find radioactive materials created by a single test.[36] Moreover, another witness testified that the seismic signals of nuclear tests could be concealed by such means as testing inside huge underground cavities.[37] In that case, there would be no suspicion that a test had occurred and no basis for conducting an inspection.[38]

On August 30, 1961, while the tripartite talks were underway, the Soviet Union withdrew from its moratorium. It pointed out that France had started testing during the moratorium, and had continued despite several warnings that French testing could lead the Soviet Union to resume testing. The Soviet statement also stressed the lack of progress in the test-ban negotiations and the rearming of West Germany.[39] It then began a massive series of nuclear tests, 50 in 1961 and 44 in 1962, that had obviously been planned well in advance.[40] One of these tests was the largest nuclear test ever conducted, with an explosive yield of 58 megatons.

The United States viewed this Soviet policy as treacherous. The New York Times editorialized that the Soviet resumption of testing "shocks the world and edges it closer to the brink of atomic holocaust."[41] A Kennedy Administration statement called it "primarily a form of atomic blackmail, designed to substitute terror for reason in the present international scene."[42] To this day, test-ban opponents cite the Soviet resumption of testing as a prime example of Soviet perfidy.

The United States, caught by surprise, responded in 1961 with a hastily planned series of ten low-yield underground tests, all of which leaked some radioactive material.[43] The United States quickly caught up with the Soviet Union in numbers of tests. In April 1962, President Kennedy ordered the resumption of atmospheric testing, and during 1962 the United States conducted 98 tests (including two for the United Kingdom), its highest annual number of tests ever. These U.S. and Soviet tests riveted public attention on the fallout issue again.

The crises in Berlin, Laos, and especially Cuba focused world attention on ways to avert nuclear war. The Hot Line agreement of 1963, in which both sides installed teletype machines in each other's capital, was one method. A nuclear test ban provided a more significant opportunity for both sides.

The individual determination of Kennedy and Khrushchev to avert war and to move U.S.-Soviet relations away from confrontation was in substantial measure responsible for obtaining agreement on a test ban. In the spring of 1963, Kennedy made a major push for a test ban. Among other things, he proposed sending a small delegation to Moscow to discuss test-ban issues with Khrushchev, and Khrushchev agreed to receive it. According to Arthur Schlesinger, Jr., "I think that Kennedy saw the main point of the treaty as a means of moving toward his . . . goal of stabilizing the international equilibrium of power."[44] Other goals for Kennedy were ending fallout and curtailing nuclear proliferation.[45] The value of a test ban for limiting weapons development was apparently of less concern.

At issue in mid-1963 was whether the test ban should be comprehensive or limited. Until June 1963, the Soviet Union had refused to consider a limited test ban without a moratorium.[46] On the other hand, a limited ban was politic-ally easier to attain given the impasse over monitoring under-ground testing. For example, in May 1963 it appeared that less than two-thirds of the Senate would support a CTBT with seven inspections annually, but that there was more support for a resolution by Senators Dodd and Humphrey calling for a ban on atmospheric and underwater tests.[47] Nonetheless, in June 1963 Kennedy declared that negotiations would begin soon in Moscow "looking forward toward early agreement on a comprehensive test ban treaty." He also indicated that the United States would cease atmospheric nuclear tests as long as others did so.[48] The U.S. position, decided on June 14, was to seek a comprehensive treaty if possible, and a limited treaty if not.[49] On the same day, Khrushchev indicated that the Soviet Union would not permit on-site inspections in a test ban treaty, and on July 2, said that the Soviet Union was willing to negotiate an agreement limiting testing in the atmosphere, under water, and in space. Negotiations began in Moscow on July 15, between the United States, the United Kingdom, and the Soviet Union.

The issue of a comprehensive ban was settled on the first day of negotiations when Khrushchev made clear that he opposed such an agreement.[50] Accordingly, the negotiators left underground testing in abeyance and focused on banning tests in the atmosphere, in space, and under water. The negotiators incorporated these elements into the Limited Test Ban Treaty (LTBT), initialed on July 25.

At the insistence of the Joint Chiefs of Staff, and to secure the Senate's advice and consent to ratification, Kennedy promised several "safeguards," notably that the United States would conduct a vigorous underground testing program, be ready to resume atmospheric testing, maintain the nuclear weapons laboratories, and improve U.S. verification capabilities. On September 24, the Senate gave its advice and consent to ratification; the treaty entered into force on October 10, 1963.

The timing and form of the LTBT were largely driven by goals and events beyond nuclear testing. A test ban of some sort was a logical agreement at the time for reducing the risk of war, slowing proliferation, and stopping fallout. It was highly visible. Both sides had extensive experience negotiating test bans, and both knew which aspects of a test ban were negotiable. Regarding the choice among test bans, expediency argued for a limited rather than a comprehensive ban. A limited ban could symbolize a step back from the brink nearly as well as a comprehensive ban. Much public pressure for a test ban came from the desire to stop fallout: a ban on atmospheric testing met this goal. While a CTBT was seen as more useful than a limited ban for slowing nuclear proliferation, the three original signatories of the LTBT tried to increase the value of the treaty for limiting proliferation by including in it a pledge to work toward a CTBT. Political opposition was much stronger to a CTBT than to a limited treaty. The technical feasibility of monitoring underground testing was uncertain. Finally, both sides wanted an agreement, and both knew that a limited treaty could be achieved quickly while a CTBT could not.

Ironically, while the LTBT pursued goals and responded to events that were largely beyond nuclear testing, the treaty dramatically changed the test ban debate. The treaty, by virtually eliminating concern over fallout, greatly reduced public pressure for subsequent test ban treaties. Because of the "safeguards," it did little to reduce concerns about the development of new weapons that testing opponents held prior to the treaty. Weapons designers were surprised by how much they could do with underground testing. Before the LTBT entered into force, they had little experience with larger underground nuclear tests for weapon development.[51] Afterwards, they used underground testing effectively for that purpose, as evidenced by the deployment of many new strategic and tactical warheads since then. At the same time, the safeguards ensured continued political and financial support for efforts to improve the technical ability to monitor test bans, notably the Department of Defense's Vela program, which started in 1959. Among other things, Vela pursued the use of seismic means to reduce the threshold at which underground nuclear explosions could be detected and identified, and the use of satellites to detect nuclear explosions in the atmosphere and in space.[52] The resultant scientific progress has been substantial, facilitating further test-ban agreements.

## 1964-1976: DETENTE

### Nuclear Concerns--Proliferation, Reduction of Permitted Test Yields

The international context of much of this period was paradoxical: U.S.-Soviet relations did not worsen despite the war in Vietnam. In fact, this period of "detente" saw many arms control measures that sought to limit the spread or increase of nuclear arms. Some measures were quite limited, such as the Outer Space Treaty and the Seabed Arms Control Treaty. Concerns about the international spread of nuclear weapons were addressed in the Treaty of Tlatelolco and the Nuclear Nonproliferation Treaty (NPT). The former seeks to ban nuclear weapons from Latin America. The latter seeks to prevent the spread of nuclear weapons to non-nuclear nations. In it, the United States, United Kingdom, and Soviet Union committed themselves again to work toward a cessation of the nuclear arms race and nuclear testing. Nonetheless, the other two nuclear-weapon states--China and France--and several near-nuclear states did not ratify the NPT, leaving nuclear nonproliferation a continuing concern.

In 1972, the two SALT I agreements were completed and entered into force. One, the Interim Agreement, limited numbers of ballistic missiles.[53] The other, the ABM Treaty, placed many constraints on the development of ballistic missile defenses and, together with a 1974 protocol, limited each side to deploying 100 interceptor missiles at one site.[54]

It is widely believed that shortly before the Watergate affair forced him from office, President Nixon pursued an arms control agreement with added urgency for political reasons.[55] When it became obvious that the second round of SALT negotiations would not produce rapid agreement, the Threshold Test Ban Treaty (TTBT)[56] was quickly negotiated; it was signed in July 1974. This treaty prohibited underground nuclear weapons tests having an explosive yield greater than 150 kt. To prevent either side from evading the 150-kt threshold under the guise of peaceful nuclear explosions, the companion 1976 Peaceful Nuclear Explosions Treaty (PNET)[67] extended this limit to tests of nuclear explosions for peaceful purposes. Both sides began observing the TTBT on March 31, 1976; the TTBT and PNET were sent to the Senate for advice and consent to ratification on July 29, 1976.

The 150-kt threshold has permitted significant weapons development. While 150-kt tests can be readily detected, assuring compliance with the threshold is harder because it requires measuring the yield of the other side's nuclear explosions. To aid in this effort, the treaties contain elaborate provisions which, as discussed later, the Reagan Administration views as inadequate.

A TTBT was supported by some opponents of testing. Those who saw it as a significant step toward a CTBT argued that the threshold could be reduced as confidence in verification increased, and that the two-year delay in implementing the treaty allowed time for high-yield tests, meeting a concern of CTBT opponents. TTBT opponents fell into two groups: (1) Supporters of unlimited underground testing predicted that the 150-kt limit would jeopardize national security. (2) Some testing opponents held that the threshold would do little or nothing to slow weapons development and was far above the level of confident verification. In the view of these testing opponents, the two-year delay allowed too much time for testing high-yield weapons. They saw the TTBT, like the Limited Test Ban Treaty before it, as a step backward for CTBT prospects. By giving the illusion of progress, they felt it reduced public pressure for a CTBT while avoiding the central goal of a CTBT, a halt to nuclear weapons development.

Neither the TTBT nor the PNET have been ratified. Ratification was deferred by President Ford during the 1976 election season, by President Carter in his attempt to reach a CTBT and SALT II, and by President Reagan because of his concern that the United States could not confidently monitor compliance with the 150-kt threshold.[58]

In retrospect, the TTBT is far removed from a CTBT. It showed one way to approach a CTBT, through declining yields. This approach, however, leaves the problem of determining whether a test complies with a threshold, a problem the CTBT does not pose. While some testing supporters feel that the two-year delay in implementing the TTBT was insufficient, it seems clear that the 150-kt threshold has not jeopardized national security. Weapons development has proceeded rapidly; indeed, one lesson of relevance to future test ban efforts is that weapons designers adapt to the limits placed on them and are able to do more within those limits than most people expect.

## 1977-1980: ARMS CONTROL INITIATIVES

### Nuclear Concern--Stockpile Confidence

A CTBT was one of President Carter's arms control goals. Soon after his inauguration, he said, "I am in favor of eliminating the testing of all nuclear devices instantly and completely."[59] In March 1977, the United States and Soviet Union agreed to resume CTBT negotiations and, as in the earlier test ban negotiations, the United Kingdom also participated. The negotiations quickly made progress. Both sides made concessions: (1) Before the negotiations began, the Soviet Union indicated at the United Nations that it would accept an unlimited number of "voluntary" or "challenge" on-site inspections, in which one side could propose to conduct an inspection of a suspicious event. The United States accepted this concept, with the understandings that the treaty contain procedures for conducting inspections and that a rejection would be viewed as a serious matter. This marked a change in U.S. policy dating from the 1950s that inspections would have to be mandatory.[60] (2) The Soviet Union withdrew its earlier position that a CTBT would have to include all nuclear powers, including China and France.[61] The Soviet side nonetheless held that continued testing by other nations could lead to a reconsideration of the treaty. (3) The Soviet Union changed its position on peaceful nuclear explosions from insisting that they be permitted under a CTBT to accepting a moratorium on these explosions for the period of the CTBT.[62] (4) The Soviet Union accepted ten unmanned seismic stations on Soviet territory, as proposed by the United States, if the United States and the United Kingdom would do the same. The United States accepted ten; the United Kingdom did not, on grounds that fewer stations were needed to monitor U.K. testing because its territory is much smaller than that of the other two nations.[63]

Nonetheless, the negotiations failed to achieve a CTBT by the end of Carter's term. Opposition was strong within the Administration. According to Herbert York, the chief U.S. CTBT negotiator during 1979 and 1980, there was opposition from Secretary of Energy James Schlesinger, the Joint Chiefs of Staff, the director of the Defense Nuclear Agency, National Security Adviser Zbigniew Brzezinski, the directors of Los Alamos and Livermore National Laboratories, many in the Department of Energy, the armed services, the National Security Council staff, and others.[64]

Arguments opposing a CTBT were presented to Congress. In testimony before the House Armed Services Committee in 1978, Department of Energy and Defense Nuclear Agency representatives emphasized that a CTBT would prevent the United States from maintaining confidence in its existing stockpile of nuclear warheads.[65] While this argument had been raised during a 1971 hearing exploring the prospects for a CTBT,[66] it had greater force in 1978 when a CTBT appeared imminent. Testimony also questioned whether the weapon laboratories could retain competent personnel under a CTBT.

York notes that the opposition had two effects.[67] First, it led Carter to switch his objective from a permanent to a temporary test ban.[68] To this end, he signed Presidential Decision Memorandum 38 setting a five-year ban as the goal of the negotiations.[69] Second, it led him to conclude that he had to bring SALT II, which restricted strategic offensive forces, to the Senate before a CTBT. The Administration was negotiating both treaties, but placed higher priority on SALT II than on a CTBT. Carter feared that if the Senate acted on a CTBT before SALT II, he might lose both.

SALT II was signed and submitted to the Senate in June 1979. But the furor in August through October 1979 over the presence of Soviet combat troops in Cuba delayed action on SALT II, and the Soviet invasion of Afghanistan in December 1979 forced the Administration to suspend its effort to secure ratification of SALT II. These events, along with the Iran hostage crisis and the 1980 presidential election campaign, made it impossible to conclude a CTBT.

## 1981-1987: COMPETITION

### Nuclear Concern--New Weapons

The past few years have seen more debate on test bans than at any time since 1963. Initially, the Reagan Administration did not address the test ban issue. Then in July 1982, President Reagan decided to suspend negotiations on a CTBT indefinitely and sought to renegotiate the TTBT and PNET.[70] He viewed the CTBT as a long-term policy goal that was not in the national interest at the time. The Administration view is summarized in a 1986 briefing by the Department of Energy:

> . . . a CTB . . . is acceptable only under certain circumstances. To be specific, a CTB must be viewed in the context of a time when we do not need to depend on nuclear deterrence to ensure international security and stability, and when we have achieved broad, deep, and verifiable arms reductions, substantially improved verification capabilities, expanded confidence-building measures, and greater balance in conventional forces.[71]

Administration and laboratory officials also point out that nuclear tests are needed for development of new warheads, maintenance of stockpile confidence, and evaluation of nuclear weapon effects.[72] Paul Brown, Assistant Associate Director for Arms Control, Lawrence Livermore National Laboratory, wrote in 1984: "In my view, as long as our national security interests depend on the deterrence of war with nuclear forces, the discontinuation of testing (i.e., a comprehensive test ban) is not consistent with those interests."[73]

Beyond that, the Administration believes the United States cannot effectively monitor Soviet compliance with a threshold. As President Reagan indicated in January 1987:

> ... the TTBT and PNET are not effectively verifiable in their present form. Large uncertainties are present in the current method employed by the United States to estimate Soviet test yields. ... effective verification of the TTBT and PNET requires that we reduce the current unacceptable level of uncertainty in our estimates of the yields of nuclear tests. ... we require ... verification through direct, on-site hydrodynamic yield (CORRTEX) measurement of all appropriate high-yield nuclear detonations.[74]

CORRTEX involves placing a cable in a shaft within 50 feet of the shaft containing a nuclear device.[75] The blast crushes the cable, and yield can be inferred from the speed at which the cable is crushed. In the Administration view, CORRTEX is useful because it is more accurate than seismic means, reducing the degree by which the Soviet Union could conduct tests exceeding the 150-kt threshold, and because it does not reveal information about warhead design to the party using it, so that the United States would not object if the Soviet Union used CORRTEX in monitoring U.S. compliance with the threshold. Accordingly, the Administration proposed renegotiating the TTBT and PNET to enhance verification, using CORRTEX on every test of 75 kilotons or more.[76]

The Soviet Union has responded in several ways. In October 1982, it submitted a document on testing to the U.N. General Assembly that "was essentially an outline of the draft treaty that had been negotiated in the trilateral negotiations during the Carter Administration."[77] The draft treaty proposed banning nuclear tests, included a moratorium on peaceful nuclear explosions, and provided that one side could request an on-site inspection, which the other would permit or provide a reason for refusing.[78] Regarding the desire to renegotiate the TTBT and PNET, the Soviet Union noted that the treaties had been signed, asserted that their verification provisions were adequate, and claimed that "the uncertainties mentioned by the U.S. side would not have taken place if the entire verification system established by the treaties had been put into effect."[79]

To dramatize its support of a CTBT, the Soviet Union conducted a unilateral moratorium on nuclear tests from August 1985 to February 1987.

Regarding the Soviet contention on the adequacy of monitoring, the Administration charged in a report of January 1984 to Congress that the Soviet Union "is likely to have violated the nuclear testing yield limit of the Threshold Test Ban Treaty."[80] Many U.S. seismologists who are involved in this issue assert that this claim is poorly supported by the data. (See p. 191.)

Congress pushed the Administration on behalf of test-ban treaties. Congress has held several hearings on test bans in the last few years. Members have introduced many bills, resolutions, and amendments urging the President to request Senate advice and consent to ratification of the TTBT and PNET and to resume negotiations toward a CTBT. In August 1986, the House amended the FY87 defense authorization bill to prevent tests exceeding one kiloton during 1987 if the Soviet Union does not conduct such tests and permits the United States to install seismic monitoring stations on its territory on a reciprocal basis.[81]

President Reagan indicated that he would veto the defense bill if it contained the test-ban provision, among others. With the announcement of the U.S.-Soviet summit to be held in Reykjavik, Iceland, in October 1986, President Reagan argued that the provision would undercut his negotiating position. Conferees on the defense bill dropped it in return for the President's agreement to make ratification of the TTBT and PNET a first order of business in the next Congress if the Soviet Union would agree to certain verification measures by early 1987. Otherwise, the President would request Senate advice and consent but would not let the treaties take effect until they are adequately verifiable. The President also agreed that after the verification concerns are met and the treaties ratified, the two nations should negotiate step-by-step reductions in nuclear testing leading eventually to a CTBT, in conjunction with a program to reduce and ultimately eliminate nuclear weapons.[82]

At Reykjavik, the Soviet Union agreed to negotiate improved verification measures, then to negotiate reductions in nuclear testing. The summit did not, however, reach agreement on actual verification improvements. Accordingly, in January 1987, the President requested the advice and consent of the Senate to ratification of the PNET and TTBT with the condition that he would not proceed with ratification until provisions were negotiated rendering the treaties effectively verifiable.[83]

Following Reykjavik, the two nations have held discussions at Geneva seeking to upgrade TTBT and PNET verification measures sufficiently for the United States to ratify these treaties. In response to the U.S. concern on verification, the Soviet Union proposed in April 1987 that each side explode one of its own nuclear devices underground at the other side's test site for calibrating seismometers.[84] In May, the Soviet Union indicated that the United States could use CORRTEX to measure the yield of some Soviet tests if the United States agrees to start negotiations on further test limits.[85] In June, the Soviet Union submitted to the Conference on Disarmament a draft CTBT that contained provisions for the announcement of test ranges, an international network for transmitting seismic data, and mandatory on-site inspections using international inspectors.[86]

Congressional action on test bans continued in 1987. In April, Senator Hatfield introduced a bill that would ban underground nuclear tests with yields over one kiloton for two years, excepting two tests with yields of 15 kilotons or less during that time, as long as the Soviet Union observes the same limitations.[87] In May, the House passed an amendment to the FY88-FY89 defense authorization bill on testing similar to the one it passed on testing in 1986.[88] Senator Hatfield's bill was converted into an amendment to the FY88-FY89 defense authorization bill; the amendment was tabled in September. The final legislation contained no provisions limiting nuclear testing.

## Conclusion: Lessons for Future Test Bans

The history of the test ban issue holds a number of lessons for consideration of more restrictive test bans.

Most striking, progress on test bans has occurred when the Soviet leader and, especially, the U.S. President personally wanted a test ban, particularly when it would promote goals beyond those stemming directly from a reduction in testing. In 1963, Kennedy and Khrushchev wanted a test ban to symbolize a move away from confrontation; for Kennedy and perhaps for Khrushchev, a test ban was also a means of stopping fallout and slowing nuclear proliferation. On the basis of earlier negotiations, it was apparent that a limited test ban treaty could be achieved rapidly, and it was, in 11 days. In 1974, President Nixon wanted a TTBT, apparently in part as a means of visibly exercising leadership and deflecting attention from the Watergate crisis. A TTBT was relatively noncontroversial and, like the 1963 treaty, was negotiated rapidly. Negotiations for a PNET continued under President Ford, and were successfully concluded, but he did not seek the Senate's advice and consent to ratification of the PNET and TTBT, reportedly to avoid potential damage to his presidential campaign. President Carter wanted a CTBT, and negotiations made rapid progress, but when he realized the extent of opposition to a CTBT and decided to pursue ratification of SALT II before a CTBT, progress toward a CTBT stopped. President Reagan views a CTBT as a long-term goal of U.S. policy, appropriate for a time when nuclear weapons are no longer needed for deterrence. Consequently, he has suspended CTBT negotiations.

While congressional opinion of most presidential test ban initiatives over the years has been mixed, with some Members supporting them and others opposing them, Congress has been the principal national forum for the opposition to these initiatives. When the Administration supported a test ban, as in 1963, Members opposed to a test ban publicized their views through hearings, reports, legislation, and floor debate. Following the LTBT, interest in test bans fell sharply. During the Nixon years, however, some Members tried to bring the test-ban issue back to public and congressional attention through hearings and resolutions. Under Reagan, Congress has been the principal national forum for supporters of the TTBT, PNET, and CTBT, using methods similar to those that anti-LTBT Members used in 1963. Members have used legislation with particular effect, with President Reagan promising in 1986 to seek Senate advice and consent to ratification of the TTBT and PNET, once adequately verifiable, and then to begin negotiations for step-by-step reductions in testing, in exchange for agreement by conferees on the FY87 defense authorization bill to drop the House amendment restricting testing. Of course, Members supporting the Administration position have at their disposal the same methods as their opponents, but the Administration is usually the most visible advocate of its own position.

In addition, Congress translates technical limits into political ones. In 1963, there was wide support in the Senate for a ban on testing in the atmosphere and under water; in fact, Senator Dodd, long an opponent of a CTBT, sponsored a resolution supporting such a ban along with Senator Humphrey, a long-time CTBT supporter. At the same time, it appeared that less than two-thirds of the Senate would support a CTBT. In 1986 and 1987, the House passed amendments banning nuclear tests above one kiloton for a year, provided the Soviet Union ceased such tests and permitted certain verification measures. The low threshold was chosen instead of a ban on all tests on grounds that one kiloton was roughly the lower limit of capability to monitor clandestine testing.[89]

The technology of verification has improved dramatically over the past three decades and continues to improve. In the late 1950s, there was little experience with detecting underground nuclear tests, seismic instrumentation was much less sophisticated, there was little understanding of the differences in seismic signals resulting from earthquakes and explosions, detection of underground explosions was possible down to about 20 kilotons with no evasion, there was concern that evasive techniques could boost the yield threshold of detection manyfold, and there was a determination to secure on-site inspections despite their low value.

Three decades of effort in monitoring test bans have yielded great progress. There is a detailed understanding of the seismology of earthquakes and explosions. Seismometers are more capable, their deployment in arrays around the world has enhanced seismic detection capability, and better data processing techniques have led to still more improvement. Nonseismic means of monitoring nuclear tests, such as CORRTEX, reconnaissance satellites, and radar, complement seismic monitoring and increase the risk of cheating.

The improved verification capability has several consequences for a more restrictive test ban. (1) The United States can, on the basis of seismic data, discriminate between earthquakes and explosions at much lower yields than it could before. (2) This enhanced capability can discriminate between earthquakes and explosions at low yields even in the face of evasive techniques such as testing in underground cavities or during an earthquake. (See pages 180, 183-184.) (3) On-site inspections may retain some political value as a deterrent to cheating, and some technical value to resolve ambiguous events. Nonetheless, emphasis has shifted from them to unmanned seismic observatories, which are less intrusive, more effective, and needed for the advances noted in points 1 and 2.

Improvements in verification technology may have influenced each side to change its political position on a more restrictive test ban. The Soviet Union has moved over the years to permit more inspections, unmanned seismic stations, and other forms of intrusive monitoring for proposed CTBTs. Technical progress may have made these concessions feasible: with the advent and refinement of reconnaissance satellites and other monitoring technologies, on-site inspections pose less risk of intelligence loss to the Soviet Union than they would have 30 years ago. At the same time, as the ability to monitor nuclear tests improved, the United States required increasingly stringent monitoring for a more restrictive test ban. Finally, U.S. desire to continue testing led to a decision not to have a CTBT in the near term regardless of verification capability. This is a marked drop in salience for the verification issue as compared to the period 1958 through 1978, when adequacy of verification seemed almost a sufficient condition for U.S. agreement to a test ban.

The salience of other test-ban arguments has also shifted over time. The development of the hydrogen bomb, the discovery that fallout can be lethal at long range, and large-scale atmospheric testing made fallout the dominant test-ban concern from the mid-1950s to 1963. The LTBT and the sharp reduction in atmospheric testing in turn reduced public concern over nuclear testing. Progress in verification technology has rendered verification less compelling as the primary argument against further restrictions on testing. As a result, supporters of continued testing have emphasized the stockpile confidence argument and the need to develop new weapons.

The decades of test-ban negotiations have paved the way for future agreements limiting nuclear testing by illuminating areas of agreement and disagreement between the two sides, indicating each side's minimum requirements for agreement, and leading each side to improve means of monitoring a test ban. This experience has led supporters and opponents of a test ban in the United States, and doubtless in the Soviet Union, to hone their arguments. Thus, if future U.S. and Soviet leaders choose to press for a more restrictive test ban, they should be able to negotiate it quickly, and both nations would have a basis for deciding expeditiously whether or not to ratify it.

## Notes

1. For histories of test bans, see Stanford Arms Control Group. International Arms Control: Issues and Agreements. Stanford, CA, Stanford University Press, 1976; and U.S. National Academy of Sciences. Nuclear Arms Control: Background and Issues. Washington, National Academy Press, 1985. For a definitive history of the events leading to the Limited Test Ban Treaty of 1963, see Jacobson, Harold Karan and Eric Stein. Diplomats, Scientists, and Politicians: The United States and the Nuclear Test Ban Negotiations. Ann Arbor, University of Michigan Press, 1966. Seaborg, Glenn. Kennedy, Khrushchev, and the Test Ban. Berkeley, University of California Press, 1981, covers the history of the test ban in the Kennedy administration. Divine, Robert. Blowing on the Wind: The Nuclear Test Ban Debate, 1954-1960. New York, Oxford University Press, 1978, covers the test ban to 1960, focusing on fallout.

2. For the speech presenting this plan, see: The Baruch Plan: Statement by the United States Representative (Baruch) to the United Nations Atomic Energy Commission, June 14, 1946. In U.S. Department of

State. Documents on Disarmament, 1945-1959, vol. I, 1945-1956. Washington, U.S. GPO, 1960. p. 7-16.

3. For the speech presenting this plan, see: Address by the Soviet Representative (Gromyko) to the United Nations Atomic Energy Commission, June 19, 1946. Ibid. p. 17-24.

4. U.S. Congress. Congressional Record. 79th Cong., 2d Sess., March 29, 1946: 2791 (Senators Huffman and Lucas, S.Res. 248); and April 18, 1946: 4023 (Representative Ludlow, H.Con.Res. 146). The Senate resolution was tabled; the House resolution was referred to the Committee on Foreign Affairs and received no further action.

5. Lepper, Mary Milling. Foreign Policy Formulation: A Case Study of the Nuclear Test Ban Treaty of 1963. Columbus, OH, Merrill, 1971. p. 25.

6. U.S. Department of Energy. Nevada Operations Office. Office of Public Affairs. Announced United States Nuclear Tests, July 1945 through December 1986. NVO-209 (rev. 7), January 1987. p. 1.

7. Cochran, Thomas, William Arkin, and Milton Hoenig. Nuclear Weapons Databook. Volume I: U.S. Nuclear Forces and Capabilities. Cambridge, MA, Ballinger, 1984. p. 11. For a discussion of the differences between types of nuclear warheads, see pp. 51-55, and especially p. 51.

8. The first successful test, Mike, of a thermonuclear device occurred in October 1952. It was not a usable weapon: it weighed 62 tons. Ibid., p. 26.

9. Glasstone, Samuel, and Philip Dolan, ed. The Effects of Nuclear Weapons, third edition. Prepared and published by the U.S. Department of Defense and the U.S. Department of Energy. Washington, U.S. GPO, 1977. p. 436-439.

10. Ibid., p. 437; and Divine, Blowing on the Wind, pp. 3-5.

11. Seaborg. Kennedy, Khrushchev, and the Test Ban, p. 4.

12. Facts on File, 1956. New York, Facts on File, Inc., 1956. p. 340.

13. U.S. Congress. Senate. Committee on Foreign Relations. Subcommittee on Disarmament. Control and Reduction of Armaments. Hearings pursuant to S. Res. 286, 84th Congress, and S. Res. 61 and 241, 85th Congress. 85th Cong., 2d Sess. Washington, U.S. GPO, 1956-1958. Parts 1-17 and index.

14. Bechhoefer, Bernhard. Postwar Negotiations for Arms Control. Washington, Brookings, 1961. p. 290-294. For a discussion of the Soviet partial measures, see ibid., p. 308.

15. Ibid., p. 294-295.

16. Ibid., p. 291.

17. Jacobson and Stein, Diplomats, Scientists, and Politicians, p. 15.

18. Bechhoefer, Postwar Negotiations for Arms Control, p. 297-308.

19. Ibid., p. 310-311.

20. Regarding the latter point, tremendous progress had been made from 1950 to 1955 in the development of nuclear weapons, notably the advent of thermonuclear weapons, boosting of fission weapons (see Chapter III), greatly improved yield-to-weight ratios, and warheads tailored to new missions. See Cochran et al., Nuclear Weapons Databook, p. 6, 11.

21. Jacobson and Stein, Diplomats, Scientists, and Politicians, p. 16. For more on the London conference, see ibid., p. 14-18.

22. Bechhoefer, Postwar Negotiations for Arms Control, p. 274-276.

23. Information in this paragraph is from the New York Times Index, 1958. New York, The New York Times Company, 1959. p. 87-88.

24. Jacobson and Stein, Diplomats, Scientists, and Politicians, p. 45-46.

25. Jacobson and Stein, Diplomats, Scientists, and Politicians, p. 49.

26. Letter from President Eisenhower to the Soviet Premier (Khrushchev), April 28, 1958. Documents on Disarmament, 1945-1959, vol. II, p. 1007.

27. U.S. Department of State. Bulletin, June 30, 1958: 1085. Cited in Jacobson and Stein, Diplomats, Scientists, and Politicians, p. 64.

28. Jacobson and Stein, Diplomats, Scientists, and Politicians, p. 80.

29. Jacobson and Stein, Diplomats, Scientists, and Politicians, p. 79.

30. Ibid., p. 136, 148-149.

31. National Academy of Sciences, Nuclear Arms Control, p. 190.

32. Jacobson and Stein, Diplomats, Scientists, and Politicians, p. 199, 236, 441.

33. U.S. Congress. Joint Committee on Atomic Energy. Special Subcommittee on Radiation and Subcommittee on Research and Development. Technical Aspects of Detection and Inspection Controls of a Nuclear Weapons Test Ban. Hearings, 86th Cong., 2d Sess., Part 1, April 1960. Washington, U.S. GPO, 1960. 478 p.

34. Ibid., p. 193. Statement of Richard Roberts, Carnegie Institution.

35. Ibid., p. 114, statement of Richard Latter, The Rand Corporation. Note that the number of unidentified seismic events at any specified threshold decreases over time as seismic detection capability improves.

36. Ibid., p. 282-305. Statement of Richard Foose, Department of Earth Sciences, Stanford Research Institute.

37. Ibid., p. 124-138. Statement of Albert Latter, The Rand Corporation.

38. Ibid., p. 288. Statement of Richard Foose.

39. Statement by the Soviet Government on the Resumption of Nuclear Weapons Tests, August 30, 1961. In Documents on Disarmament, 1961, p. 337-348.

40. "Experts estimate [that the test series] must have taken a minimum of six months, and more likely a year or more to prepare." Jacobson and Stein, Diplomats, Scientists, and Politicians, p. 280-281.

41. Soviet Policy of Terror. New York Times, Sept. 1, 1961: 16.

42. U.S. White House. Statement on the Resumption of Soviet Tests, August 31, 1961. Reprinted in U.S. Arms Control and Disarmament Agency. Documents on Disarmament, 1961. Washington, U.S. Govt. Print. Off., 1962, p. 350.

43. See Department of Energy, Announced United States Nuclear Tests, p. 10, for data on these tests.

44. Schlesinger, Jr., Arthur. A Thousand Days: John F. Kennedy in the White House. New York, Fawcett Premier, 1965. p. 831.

45. Ibid., p. 421, 819.

46. Jacobson and Stein, Diplomats, Scientists, and Politicians, p. 454.

47. Ibid., p. 447-449.

48. Kennedy, John F. Address at American University, June 10, 1963. In U.S. Arms Control and Disarmament Agency. Documents on Disarmament, 1963. Washington, U.S. GPO, 1964. p. 215-222.

49. Information in this paragraph is based on Jacobson and Stein, Diplomats, Scientists, and Politicians, p. 449-456. For text of the Treaty Banning Nuclear Weapon Tests in the Atmosphere, in Outer Space and Under Water, signatories, and discussion, see U.S. Arms Control and Disarmament Agency. Arms Control and Disarmament Agreements: Texts and Histories of Negotiations. Washington, U.S. GPO, 1982. p. 34-47.

50. Seaborg. Kennedy, Khrushchev, and the Test Ban, p. 240-241.

51. Before the 1958 moratorium, the United States conducted only 22 of 196 announced nuclear explosions underground, of which all but one yielded 5 KT or less. After the moratorium, the United States conducted many announced underground tests, but only seven of those conducted before the treaty entered into force yielded more than 20 KT. Department of Energy, Announced United States Nuclear Tests, p. 1-17.

52. For a discussion of the role of Vela in fostering progress in the seismology of monitoring nuclear test bans, see Ann Kerr, ed. The VELA Program: A Twenty-Five Year Review of Basic Research. Washington, Defense Advanced Research Projects Agency, 1985. 964 p.

53. Interim Agreement Between the United states of America and the Union of Soviet Socialist Republics on Certain Measures with Respect to the Limitation of Strategic Offensive Arms. Entered into force October 3, 1972. For text, understandings, and discussion, see Arms Control and Disarmament Agreements, p. 132-136, 148-157.

54. Treaty Between the United States of America and the Union of Soviet Socialist Republics on the Limitation of Anti-Ballistic Missile Systems. Entered into force October 3, 1972. For text, understandings, and discussion, see Arms Control and Disarmament Agreements: 132-147. Protocol to the Treaty Between the United States of America and the Union of Soviet Socialist Republics on the Limitation of Anti-Ballistic Missile Systems. Entered into force May 24, 1976. For text and discussion, see ibid., p. 161-163.

55. See, for example, National Academy of Sciences, Nuclear Arms Control, p. 197.

56. Treaty Between the United States of America and the Union of Soviet Socialist Republics on the Limitation of Underground Nuclear Weapon Tests; signed at Moscow July 3, 1974. For text, protocol, and discussion, see Arms Control and Disarmament Agreements, p. 164-170.

57. Treaty Between the United States of America and the Union of Soviet Socialist Republics on Underground Nuclear Explosions for Peaceful Purposes. Signed at Washington and Moscow May 28, 1976. For text, protocol, agreed statement, and discussion, see ibid., p. 171-189.

58. Regarding the decisions by Presidents Ford and Carter not to seek ratification of the TTBT and PNET, see National Academy of Sciences, Nuclear Arms Control, p. 199.

59. Cited in Knight, Albion, Brig. Gen., U.S. Army (Ret.), The High Security Cost of CTB. Air Force Magazine, June 1978: 66.

60. National Academy of Sciences, Nuclear Arms Control, p. 201.

61. Kincade, William. Banning Nuclear Tests: Cold Feet in the Carter Administration. Bulletin of the Atomic Scientists, November 1978: 8.

62. Ibid.

63. National Academy of Sciences, Nuclear Arms Control, p. 201.

64. York, Herbert. Testimony before the California State Senate Committee on Health and Human Services. Hearing. Forum on the Involvement of the University of California in Nuclear Testing at Lawrence Livermore and Los Alamos National Laboratories, Transcript. Sacramento, CA. February 11, 1987. p. 55-58.

65. U.S. Congress. House. Committee on Armed Services. Intelligence and Military Application of Nuclear Energy Subcommittee. Current Negotiations on the Comprehensive Test Ban Treaty. Hearings, 95th Cong., 2d Sess., March 15-16, 1978. H.A.S.C. no. 95-62. Washington, U.S. GPO, 1978.

66. U.S. Congress. Senate. Committee on Foreign Relations. Subcommittee on Arms Control, International Law and Organization. Prospects for Comprehensive Nuclear Test Ban Treaty. Hearings, 92d Cong., 1st Sess. Washington, U.S. GPO, 1971. p. 101-109.

67. York, testimony before the California State Senate, p. 58-59.

68. "The argument was not really over the duration of the treaty, but how the treaty was to be extended--whether automatically, unless one or more parties object; or whether the treaty ends and can be extended only by positive action of all the parties. The former was the Soviet position; the latter, the U.S. position." Warren Heckrotte, Lawrence Livermore National Laboratory, personal correspondence, August 1987.

69. Wilson, George. Carter to Seek 5-Year Ban on Nuclear Testing. Washington Post, May 27, 1978: 17. Cited in Kincade, Banning Nuclear Tests, p. 8.

70. Miller, Judith. U.S. Confirms a Plan to Halt Talks on a Nuclear Test Ban. New York Times, July 21, 1982: 1.

71. U.S. Department of Energy. Briefing on nuclear test ban issues, c. August 1986. p. 15.

72. See, for example, response of the Arms Control and Disarmament Agency to questions submitted by Representative Mrazek. In U.S. Congress. House. Committee on Appropriations. Subcommittee on the Departments of Commerce, Justice, and State, the Judiciary, and Related Agencies. Department of Commerce, Justice, and State, the Judiciary, and Related Agencies Appropriations for 1984. Hearings, 98th Cong., 1st Sess., Part 2. Washington, U.S. GPO, 1983. p. 273.

73. Brown, Paul. Letter to the Editor. New York Times, August 8, 1984: 22.

74. Reagan, Ronald. Soviet Union-United States Nuclear Testing. Message to the Senate, January 13, 1987. U.S. National Archives and Records Administration. Office of the Federal Register. Weekly Compilation of Presidential Documents. Vol. 23, No. 2, January 19, 1987. p. 23.

75. For a detailed description of CORRTEX, see page 7-19.

76. Gordon, Michael. Soviet Offers to Allow Some On-Site Test Monitoring. New York Times, June 4, 1987: 3.

77. National Academy of Sciences, Nuclear Arms Control, p. 203.

78. Ibid.; and Institute for Defense and Disarmament Studies. The Arms Control Reporter: A Chronicle of Treaties, Negotiations, Proposals. Brookline, MA, IDDS, 1982. p. 608.B.20. A refusal would of course be a serious matter, possibly creating grounds for withdrawing from the treaty.

79. U.S.S.R. Embassy Statement on 'Threshold' Treaties. TASS, April 13, 1983. Cited in U.S. Foreign Broadcast Information Service. Daily Report, Soviet Union, April 14, 1983: AA1.

80. Cited in National Academy of Sciences. Nuclear Arms Control. p. 204. The Administration has repeated this charge in several subsequent reports. See, for example, U.S. White House. Office of the Press Secretary. [Press Release] The President's Unclassified Report on Soviet Noncompliance with Arms Control Agreements. March 10, 1987. p. 12.

81. Amendment by Representatives Aspin, Gephardt, Schroeder, and others to H.R. 4428; passed House, 234-155, on August 8, 1986. For debate, see U.S. Congress. Congressional Record (daily edition). August 8, 1987: H 5738-H 5756. For text, see ibid., p. H 5754.

82. This agreement is detailed in U.S. Congress. House. National Defense Authorization Act for Fiscal Year 1987. Conference report to accompany S. 2638. H. Rept. 99-1001. 99th Cong., 2d Sess. Washington, U.S. GPO, 1986. p. 516-517.

83. Reagan, Ronald. Soviet Union-United States Nuclear Testing. Message to the Senate, January 13, 1987. U.S. National Archives and Records Administration. Office of the Federal Register. Weekly Compilation of Presidential Documents. Vol. 23, No. 2, January 19, 1987. p. 22-24.

84. Arms Control Reporter, p. 605.B.59-60.

85. Gordon, Michael. Soviet Offers to Allow Some On-Site Test Monitoring. New York Times, June 4, 1987: 3; and Arms Control Reporter, p. 605.B.62-63.

86. Soviet Test Ban Offer Mandates Inspections. New York Times, June 10, 1987: 12; and Arms Control Reporter, p. 608.B.140.

87. S. 1106, the Underground Nuclear Explosions Control Act.

88. Amendment by Representative Schroeder to H.R. 1748, passed by House, 234-187, on May 19, 1987. For debate, see U.S. Congress. Congressional Record (daily edition), May 19, 1987: H 3686-H 3708. For text, see ibid., p. H 3702.

89. Statement of Representative Gephardt, Congressional Record (daily edition), August 8, 1986: H 5739; and statement of Representative Oakar, Congressional Record (daily edition), May 19, 1987: H 3688.

# 3

# Overview of U.S. and Soviet Nuclear Weapons and Testing Programs

*Robert Civiak*

This chapter presents basic information about the design and functioning of nuclear warheads and U.S. and Soviet nuclear weapons and nuclear weapons testing programs.

### Design and Functioning of Nuclear Warheads[1]

Two nuclear processes--fission and fusion--are responsible for the tremendous energy release of nuclear warheads. Fission is the splitting apart of heavy nuclei like uranium and plutonium. It was the sole energy source of the first generation of nuclear weapons (A-bombs), it is the energy source of nuclear reactors, and it is an essential energy source in modern nuclear weapons. Fusion is the joining together of light nuclei like deuterium and tritium, which are two forms of hydrogen. It is the energy source of the sun and stars and is used in powerful second generation nuclear weapons, called thermonuclear weapons or H-bombs.

The basic design of a modern thermonuclear warhead, or "hydrogen bomb," is shown in schematic form in Figure 1. The design shown is not the only type of nuclear warhead possible, but the vast majority of U.S. and Soviet warheads are believed to be of this type. It consists of a "primary," in which most of the explosive energy comes from nuclear fission, and a "secondary," in which additional energy comes from nuclear fusion. A fission

## Figure 1

## Physics Package of a Thermonuclear Weapon

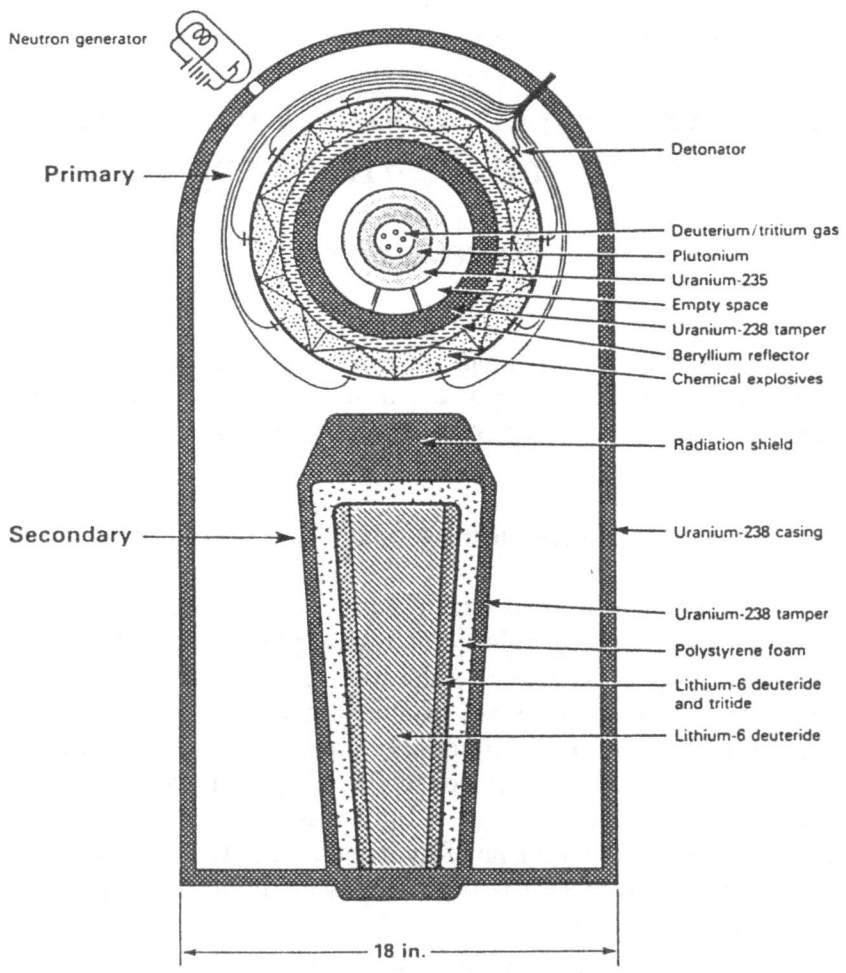

warhead, or "atomic bomb," consists only of the primary and associated casing, fuzing, etc; it obtains all its energy from fission.

Explosion of the fission primary is initiated by electronic signals sent to a number of detonators spaced around it. The detonators set off a chemical high explosive, which forms the outer portion of the primary and is specially shaped to implode or compress the inner material. At precisely the right time, a stream of neutrons is injected into the compressed primary, starting the nuclear fission chain reaction. The injected neutrons strike uranium or plutonium nuclei, causing them to split. A nucleus of uranium or plutonium, when split, produces two smaller nuclei, energy, and two or more neutrons. The new neutrons strike other nuclei and cause them to fission, thus producing more energy and more neutrons, which initiate more fissioning, and so on. Each successive generation of fissioning is larger than the previous generation. About 60 generations of fissioning occur in less than a microsecond (one millionth of a second).[2] More than half the total energy release occurs in the last generation and 99.9% of the energy release occurs in the last seven generations.[3] The yield of pure fission devices is limited by their flying apart before enough neutrons are created to cause all the fuel to fission--hence the high leverage of boosting, discussed below.

While simple in concept, the actual design of modern fission primaries is quite complex. Primaries consist of several precisely fabricated concentric shells of material including: (1) the chemical high explosive; (2) a shell of a heavy material (called the tamper) whose inertia holds the exploding mass together long enough to allow more complete fissioning; (3) a shell of materials that reflect neutrons to reduce their leakage and thus increase fissioning; (4) an empty space to allow the tamper to gather momentum before hitting the core;[4] (5) one or more shells of different types of fissionable material, i.e., different mixes of isotopes of uranium and plutonium; and (6) an inner chamber that can be filled with deuterium and tritium, which, after being compressed and heated, fuse in sufficient numbers to produce large amounts of neutrons to further boost the fissioning that occurs. The use of deuterium-tritium boosting can be controlled to vary the yield of the device.[5] By producing a large amount of neutrons at the proper time, boosting can increase the yield of a primary by many times even though the fusion energy produced by boosting is only a small fraction of the overall yield.

As noted above, most of the energy release from the primary occurs in the last few fission generations. For the warhead to function properly the compression must be highly symmetric and the material must hold together long enough for the maximum amount of fissioning to occur. Differences of a few hundredths of a microsecond can markedly change a warhead's yield. The positions and timing of the detonators, the shape and composition of the high explosive, and near perfect symmetry of the shells of tamper, reflector, and fissionable material are all critical for achieving the conditions needed for high yields.

The primary described above is suitable alone for warheads with yields up to a few tens of kilotons. However, for higher yields, the shell(s) of fissionable material would have to be made very large, which would make for an unwieldy weapon.[6] Thus, in the early 1950s, weapons designers introduced thermonuclear weapons (i.e., H-bombs), which rely upon the fusion of deuterium and tritium to greatly increase their energy release. The bulk of the fusion occurs in a part of the warhead separate from the fission primary, called the secondary. Thermonuclear weapons can have much larger yields for a given size and weight than pure fission weapons.

Temperatures in excess of 100 million degrees Celsius and very high pressures are necessary for deuterium and tritium to fuse. The explosion of a fission bomb is thus far the only way that mankind has produced the conditions needed to achieve large amounts of fusion. Deuterium and tritium are gases, which are difficult to contain in large amounts in a practical weapon. Therefore, in most thermonuclear weapons the raw fusion fuel is lithium-six deuteride ($Li^6D$),[7] a solid material that is relatively easy to handle. The lithium atoms absorb neutrons that are produced by fission reactions in the early stages of the warhead's detonation. The neutrons cause the lithium atoms to split into tritium and helium, supplying the tritium for deuterium-tritium (DT) fusion to occur. The DT fusion produces further neutrons that also convert lithium-six to tritium. While the need to breed tritium complicates the design, the hard part in designing a thermonuclear warhead is to generate the high pressures and temperatures needed in a manner that allows the desired amount of fusion to occur before the warhead is blown apart.

In addition to the components already discussed, U.S. nuclear weapons contain complex fuzing and firing mechanisms to arm and detonate the weapon, safety and control mechanisms to prevent accidental or unauthorized detonation, and a microprocessor to control all the electronic operations.[8] The fuze is a combination of sensors and electronics that controls the timing of detonation. Its design varies with the intended mission of a weapon. In a missile, for example, the fuze may start the arming and firing mechanisms at preset heights above the ground (determined, for example, by pressure measurements or by radar). The firing mechanism consists of the batteries, capacitors, and other electronics that actually fire the detonators. For safety reasons, the firing mechanism cannot work until switches activated by the proper code connect the detonators to the fire set, the weapon is armed, and the proper signal is received from the fuze. Safety mechanisms, including a device to prevent unauthorized use called a permissive action link (PAL), take on many forms. Most such devices prevent detonation of the primary by interrupting the functioning of the electronic fuzing and firing mechanisms. PALs are disengaged either by the correct manipulation of a combination lock type mechanism or by receipt of the proper coded electronic signal. Some modern PALs include intrusion sensing devices and warhead disabling features. Because these components (as well as the neutron generator and perhaps additional guidance and control systems) are all adjacent to or within the warhead's outer shell, their design and placement can affect its yield.

## U.S. Nuclear Weapons Stockpile

### *Existing Weapons*

According to the Stockholm International Peace Research Institute (SIPRI), as of 1987 the United States had approximately 13,700 strategic nuclear warheads in its stockpile and another

10,000 warheads for theater and tactical nuclear systems.[9] Warheads of about 65 distinct designs have been in the stockpile at one time or another since 1945; another 25 designs were taken to advanced stages of development but never entered the stockpile.[10] The current stockpile contains 26 distinct warhead designs with about 40 different models in all, counting modifications and different yield varieties (see Table 1).[11] The average age of the nuclear weapons in the stockpile is about 13 years.[12]

Nuclear warheads are used with at least 37 different delivery systems for a large variety of strategic and tactical missions. The diversity of U.S. warheads and delivery systems is shown in Tables 2 and 3. The smallest warhead in the U.S. stockpile weighs under 60 pounds and has a subkiloton yield. The largest warhead in the U.S. stockpile, the B53 bomb, weighs 8,850 pounds and has a yield in the megaton range. The Air Force has been replacing the B53 with the more modern, lower-yield B83, and few remain in the stockpile. It has been reported, however, that the Air Force is now removing retired B53 bombs from storage and returning them to the active stockpile aboard B-52 bombers.[13] Five warheads (seven counting modifications) are currently in production and are listed below:[14]

- B61 (mod-3 and mod-4) tactical bomb used on a variety of aircraft;
- W80 warhead for the Tomahawk sea-launched cruise missile (mod-0) and the Air Force air-launched cruise missile (mod-1);
- B83 bomb for the B-1 and other bombers;
- W84 warhead for the ground-launched cruise missile; and
- W87 warhead for the MX missile.

### Weapons Development

Despite the diversity of warhead designs in the stockpile, the United States continues to develop new nuclear warheads to take advantage of technological developments, to improve the safety

## Table 1

## U.S. Nuclear Weapons Stockpile (June 1987)

| Warhead/Weapon | First produced | Yield (kilotons) | User | Number | Status |
|---|---|---|---|---|---|
| **Bombs and bomber weapons** | | | | | |
| B28/bomb* | 8/58 | 70–1,450 | AF, NATO | 1,000 | Being replaced by new B61 and B83 bombs. |
| B43/bomb* | 4/61 | <1,000 | AF, MC, N, NATO | 975 | Being replaced by new B61-3, B61-4, and B83 bombs. |
| B53/strategic bomb* | 8/62 | 9,000 | AF | 25 | Being replaced by B83 bomb. |
| B57/bomb* | 1/63 | <1 to 20 | AF, MC, N, NATO | 1,195 | To be replaced by B61 and new strike bomb. |
| B61-0,1,7†/bomb | 10/66 | <1 to 500 | AF | 1,000 | Strategic bomb replacing B28. |
| B61-2,3†,4†,5/bomb | 3/75 | <1 to 345 | AF, MC, N, NATO | 2,150 | Tactical bomb replacing B28, B43, and B57. |
| B83†/bomb | 6/83 | low–1,200 | AF | 1,000 | Replacing strategic B28 and B43 bombs. |
| W69/SRAM (short-range attack missile) | 10/71 | 170 | AF | 1,175 | To be replaced by SRAM II. |
| **Air-defense missiles** | | | | | |
| W31/Nike Hercules* | 10/58 | <1 to 20 | NATO | 75 | U.S. missiles retired in 1984; NATO missile to be replaced by conventional Patriot. |
| W45/Terrier* | 1/62 | ~1 | N | 285 | May be replaced by Standard Missile-2 (nuclear)/W81. |
| **Artillery and demolitions** | | | | | |
| W33/8-inch artillery* | 1/57 | <1 to 12 | A, MC, NATO | 900 | A portion has been replaced by new W79. |
| W48/155mm artillery* | 10/63 | 0.1 | A, MC, NATO | 925 | To be replaced by non-enhanced radiation W82. |
| W54/special atomic demolition munition | 4/61 | 0.01–1 | A, N, MC | 300 | Used for Special Operations; no planned replacement. |
| W79/8-inch artillery (enhanced radiation) | 9/81 | 0.8 | A, MC | 40 | Will be converted to fission versions. |
| W79/8-inch artillery | 10/84 | 1.1 | A, MC, NATO | 300 | Production completed August 1986. |
| **Antisubmarine weapons** | | | | | |
| W44/ASROC* (ship-launched) | 5/61 | ~1 | N | 575 | May be replaced by antisubmarine standoff weapon. |
| W55/SUBROC* (submarine-launched) | 6/64 | 1–5 | N | 285 | Scheduled for retirement in 1989. |
| B57/depth bomb* | 1/63 | <1 to 20 | N | 900 | To be replaced by new depth bomb. |
| **Intermediate- and short-range missiles** | | | | | |
| W50/Pershing 1a | 3/63 | 60, 200, 400 | NATO | 100 | U.S. missiles were replaced by Pershing II/W85. |
| W70-0, 1, 2/Lance | 6/73 | 1–100 | A, NATO | 905 | To be replaced by tactical follow-on missile. |
| W70-3/Lance (enhanced radiation) | 5/81 | <1 | A | 380 | Stored in the U.S. |
| W85/Pershing II | 2/83 | 0.3–80 | A | 120 | Replaced U.S. Pershing 1a, Dec. 1983 to Dec. 1985. |
| **Submarine-launched ballistic missiles** | | | | | |
| W68/Poseidon C3 | 5/70 | 40 | N | 2,800 | Being retired and replaced by Trident I/W76. |
| W76/Trident I C4† | 6/78 | 100 | N | 3,000 | To be eventually replaced by Trident II D5. |
| **Intercontinental ballistic missiles** | | | | | |
| W56/Minuteman II | 3/63 | 1,200 | AF | 500 | No planned silo replacement. |
| W62/Minuteman III | 3/70 | 170 | AF | 740 | 150 warheads to be replaced by MX. |
| W78/Minuteman III | 8/79 | 335 | AF | 950 | Mk 12A retrofitted between Dec. 1979 and Feb. 1983. |
| W87/MX† | 4/86 | 300 | AF | 250 | 500 warheads planned by Dec. 1988. |
| **Cruise missiles** | | | | | |
| W80-0/Tomahawk† (sea-launched) | 12/83 | 5–150 | N | 125 | 758 missiles planned for 198 ships and submarines. |
| W80-1/ALCM† (air-launched) | 12/81 | 5–150 | AF | 1,650 | 1,500 more warheads planned for Advanced Cruise Missile. |
| W84/GLCM† (ground-launched) | 6/83 | 0.2–150 | AF | 300 | An additional 200 warheads by end of 1988. |

*Weapons scheduled in present plans for complete or partial retirement 1987–1990s.  †In production.

A: Army; AF: Air Force; MC: Marine Corps; N: Navy; NATO: non-U.S. delivery systems. In weapons nomenclature, B stands for "bomb" and W for "warhead." The number following the letter indicates the order in which it entered full-scale development; for example, W69 followed W68.

## Table 2

## U.S. Strategic Nuclear Forces, 1987

| Weapon system | No. | Year | Range | Warheads | | No. in |
| Type | deployed | deployed | (km) | Warhead × yield | Type | stockpile |
|---|---|---|---|---|---|---|
| *ICBMs*[a] | | | | | | |
| Minuteman II | 450 | 1966 | 11 300 | 1 × 1.2 Mt | W-56 | 480 |
| Minuteman III (Mk 12) | 240 | 1970 | 13 000 | 3 × 170 kt | W-62 | 750 |
| Minuteman III (Mk 12A) | 300 | 1979 | 13 000 | 3 × 335 kt | W-78 | 950 |
| MX | 10 | 1986 | 11 000 | 10 × 300 kt | W-87 | 110 |
| Total | 1 000 | | | | | 2 290 |
| | | | | | | |
| *SLBMs* | | | | | | |
| Poseidon | 256 | 1971 | 4 600 | 10 × 50 kt | W-68 | 2 750 |
| Trident I | 384 | 1979 | 7 400 | 8 × 100 kt | W-76 | 3 300 |
| Total | 640 | | | | | 6 050 |
| | | | | | | |
| *Bombers* | | | | | | |
| B-1B | 18 | 1986 | 9 800 | 8–24 | | 250 |
| B-52G/H | 263 | 1955 | 16 000 | 8–24[a] | | 4 733 |
| FB-111 | 61 | 1969 | 4 700 | 6[b] | | 360 |
| Total | 339 | | | | | 5 343 |
| | | | | | | |
| *Refuelling aircraft* | | | | | | |
| KC-135 | 615 | 1957 | .. | .. | .. | .. |

[a] The four Titan II ICBMs remaining at Dec. 1986 are scheduled to be deactivated by mid-1987.

[a] Bomber weapons include six different nuclear bomb designs (B-83, B-61-0, -1, -7, B-57, B-53, B-43, B-28) with yields from sub-kt to 9 Mt, ALCMs with selectable yields from 5 to 150 kt, and SRAMs with a yield of 200 kt. FB-111s do not carry ALCMs or B-53 or B-28 bombs.

Source: SIPRI Yearbook 1987: World Armaments and Disarmament, p. 6-7. Oxford: Oxford University Press, 1987. Copyright 1987 Stockholm International Peace Research Institute. Reproduced with the permission of Oxford University Press. All rights reserved.

## Table 3

### U.S. Theater Nuclear Forces, 1987

| Weapon system Type | No. deployed | Year deployed | Range (km) | Warheads Warhead × yield | Type | No. in stockpile |
|---|---|---|---|---|---|---|
| **Land-based systems:** | | | | | | |
| Aircraft[a] | 2 000 | .. | 1 060–2 400 | 1–3 × bombs | [*] | 2 800 |
| | | | | | | |
| *Missiles* | | | | | | |
| Pershing II | 108 | 1983 | 1 790 | 1 × 0.3–80 kt | W-85 | 125 |
| GLCM | 208 | 1983 | 2 500 | 1 × 0.2–150 kt | W-84 | 250 |
| Pershing 1a | 72 | 1962 | 740 | 1 × 60–400 kt | W-50 | 100 |
| Lance | 100 | 1972 | 125 | 1 × 1–100 kt | W-70 | 1 282 |
| Honest John | 24 | 1954 | 38 | 1 × 1–20 kt | W-31 | 132 |
| Nike Hercules | 27 | 1958 | 160 | 1 × 1–20 kt | W-31 | 75 |
| | | | | | | |
| *Other systems* | | | | | | |
| Artillery[b] | 4 300 | 1956 | 30 | 1 × 0.1–12 kt | [*] | 2 022 |
| ADM (special) | 150 | 1964 | .. | 1 × 0.01–1 kt | W-54 | 150 |
| | | | | | | |
| **Naval systems:** | | | | | | |
| *Carrier aircraft* | 900 | .. | 550–1 800 | 1–2 × bombs | [*] | 1 000 |
| | | | | | | |
| *Land-attack SLCMs* | | | | | | |
| Tomahawk | 100 | 1984 | 2 500 | 1 × 5–150 kt | W-80-0 | 110 |
| | | | | | | |
| *ASW systems* | | | | | | |
| ASROC | .. | 1961 | 10 | 1 × 5–10 kt | W-44 | 574 |
| SUBROC | .. | 1965 | 60 | 1 × 5–10 kt | W-55 | 150 |
| P-3/S-3/S11-3[c] | 630 | 1964 | 2 500 | 1 × <20 kt | B-57 | 897 |
| | | | | | | |
| *Naval SAMs* | | | | | | |
| Terrier | .. | 1956 | 35 | 1 × 1 kt | W-45 | 290 |

[a] Aircraft include Air Force F-4, F-16 and F-111, and NATO F-16, F-104 and Tornado. Bombs include four types (B-28, B-43, B-57 and B-61) with yields from sub-kt to 1.45 Mt.

[b] There are two types of nuclear artillery (155-mm and 203-mm) with four different warheads: a 0.1-kt W-48, 155-mm shell; a 1- to 12-kt W-33, 203-mm shell; a 0.8-kt W-79-1, enhanced-radiation, 203-mm shell; and a variable yield (up to 1.1 kt) W-79-0 fission warhead. The enhanced radiation warheads will be converted to standard fission weapons.

[c] Aircraft include Navy A-6, A-7, F/A-18 and Marine Corps A-4, A-6 and AV-8B. Bombs include three types with yields from 20 kt to 1 Mt.

[d] Some US B-57 nuclear depth bombs are allocated to British Nimrod, Italian Atlantique and Dutch P-3 aircraft.

Source: SIPRI Yearbook 1987: World Armaments and Disarmament, p. 6-7. Oxford: Oxford University Press, 1987. Copyright 1987 Stockholm International Peace Research Institute. Reproduced with the permission of Oxford University Press. All rights reserved.

and security of weapons systems, and to respond to changes in Soviet forces.

Technical advances over the past 40 years have enabled weapons designers to make large reductions in the size and weight of nuclear warheads. The ratio of explosive yield to warhead weight (called the yield-to-weight ratio) has been increased to more than 600 times its early values.[15] The introduction of thermonuclear weapons in the 1950s was a major step in improving yield-to-weight ratios. Most of the improvement in yield-to-weight ratios had occurred by the early 1960s. Since then, much smaller gains have occurred through miniaturization of electronics and other components (such as detonators and neutron generators) and it is unlikely that further miniaturization alone would be sufficient motivation for new warhead designs in the future.

Reducing the size and weight of warheads had major strategic significance. First, it made weapons delivery via ballistic missiles possible, thereby reducing the time for delivery from hours to minutes and making delivery highly resistant to defenses. Later, it allowed for the introduction of Multiple Independently Targetable Reentry Vehicles (MIRVs) on ballistic missiles, which greatly reduced the cost of delivering warheads to their targets. These developments also have some adverse consequences for strategic stability. Both sides now rely heavily on missiles that can reach their targets in minutes and cannot be recalled or disabled, and MIRVs have introduced the possibility of one side destroying the other side's ballistic missiles in a first strike with only a portion of its own missiles.

While radical advances in nuclear warhead technology have not occurred for many years, the technology continues to evolve with new warhead designs. The major reason for developing new warheads has been to tailor them to new missions and delivery systems by optimizing characteristics such as size, weight, explosive yield, the form of the energy released (e.g., blast, radiation, or neutrons), performance in different environments (e.g., in air or space, on the surface of the earth, under water or under ground), resistance to hostile threats, resistance to rapid acceleration or deceleration, reliability, safety, security, and ease of handling.

Examples of the tailoring of new warheads to meet the demands of new weapons systems include the warheads for the MX and Trident II missiles. Those warheads have been designed to meet stringent size, weight and yield requirements to maximize the range, accuracy, and lethality of those missiles. The warheads have also been provided with upgraded shielding and specially designed electronic components to resist the intense radiation and high neutron fluxes they might encounter from preceding nuclear explosions or Soviet defensive systems. Another warhead tailored to the requirements of a new weapons system is the B83 bomb carried by the B-1 bomber. The B-1 flies low to its target to avoid detection. Its bomb must therefore survive the shock of hitting the ground at high speed and delay detonation for the aircraft crew to escape. The B83 is the first bomb to have this capability.

The desire to introduce safety improvements rarely generates the requirement for a new warhead or a major redesign of an existing one. Instead, safety improvements are generally incorporated on new warheads developed for other reasons.[16] The most significant safety improvements have been the introduction of an insensitive chemical high explosive (IHE) and the development of PALs. It is very unlikely that a nuclear explosion would result from accidental detonation of a weapon's chemical explosive. However, detonation of chemical explosives could scatter uranium and plutonium and require extensive cleanup efforts as happened when B-52s carrying nuclear weapons crashed at Palomares, Spain, in 1966 and at Thule, Greenland, in 1968. IHE can withstand being dropped from an aircraft or hit with rifle bullets without detonation. It therefore provides assurance that even an accidental chemical explosion is unlikely. About one-third of the weapon systems, including the majority of the bombs, in the U.S. stockpile use IHE.[17] The properties of IHE are sufficiently unique that replacing old explosives with IHE is a significant design change that in general requires nuclear testing to provide sufficient confidence in the new warhead.[18]

The use of PALs to prevent unauthorized detonation of nuclear weapons has been noted above. PALs require that either a six-digit or twelve-digit code be applied before the weapon may be armed. Advanced PALs limit the number of attempts at inputting the code and have systems that can destroy key components and render a warhead unusable if an attempt is made to bypass the PAL. In some cases the PAL is an intimate part of the design and a major modification to the PAL could require the weapon to be tested to assure it would work. All U.S. nuclear weapons based outside the United States have advanced PALs. Some weapons securely based inside the United States and on submarines do not have advanced PALs because it was determined that their low security risk did not justify the additional weight of the PAL, which would reduce the effectiveness of the weapons.[19]

The development of nuclear warheads by the Department of Energy and deployment by the armed services proceeds through seven distinct phases. Most of the phases were defined in an agreement between the Atomic Energy Commission and the Department of Defense dated March 21, 1953, that is still in effect.[20]

Phase 1 -- Concept definition. Evaluation of new concepts and/or technology for future weapons.

Phase 2 -- Feasibility study. Examination of the feasibility and desirability of a new weapon and establishment of the military characteristics of the warhead (e.g. yield, size, survivability characteristics, fusing requirements)

Phase 2a -- Design definition and cost study. Generally each of the two weapons development laboratories (Los Alamos National Laboratory and Lawrence Livermore National Laboratory) will have one or more design options for a particular warhead at the beginning of phase 2a. At the end of phase 2a one option and the laboratory to develop it is selected.

Phase 3 -- Full scale development. Development engineering entailing extensive computer simulation, nuclear and nonnuclear testing, and materials development culminating in the establishment of the detailed design for the warhead. Up to phase 3, development is funded through general program appropriations. Congressional approval of line item funding is generally obtained as a warhead enters phase 3.

Phase 4 -- Production engineering. Activities focus on designing specialized machines and setting up manufacturing processes to produce warheads of the design chosen in phase 3. Phases 3 and 4 may overlap.

Phase 5 -- First production. Production and delivery of the first warhead(s) which are then evaluated by DOE and DOD. Evaluation includes a nuclear explosive "proof test" conducted on an early production warhead recalled for testing one or two years after being placed in the field.

Phase 6 -- Quantity production and stockpile. Mass production of a standardized warhead design and delivery into the active stockpile. An extensive reliability program is conducted on warheads in the stockpile and modifications are undertaken, if necessary.

Phase 7 -- Retirement. Elimination of the warhead from the stockpile and recycling of the nuclear materials.

Two warheads are currently in the production engineering phase of development (phase 4).[21] They are:

- The W88 warhead for the Trident II submarine-launched missile to increase its yield; and
- The W82 warhead for the 155 mm artillery-fired atomic projectile (AFAP), which will have better range, accuracy, and reliability than the one it replaces.

Two other warheads have been approved for full scale development (phase 3). They are:

- A warhead for the small ICBM (SICBM or Midgetman) to enhance the survivability of the U.S. land-based strategic deterrent by providing for mobility; and
- A warhead for the short-range attack missile (SRAM-II), a tactical air-to-surface missile with better range, accuracy, and safety features than its predecessor, the SRAM.

A modified version of the W87 warhead (used on the MX missile) has been selected as the warhead for the SICBM.[22] However, work on the modification is proceeding slowly pending a possible decision to cancel the SICBM.

Two warheads for antisubmarine warfare have completed phase 2a development, but the Department of Defense has not authorized proceeding to phase 3. They are:

- An antisubmarine warfare nuclear depth/strike bomb (ASW ND/SB), with dual capability to be used against submarines and surface targets; and
- An antisubmarine warfare standoff weapon (ASW SOW), also called Sea Lance, to replace the current SUBROC missile. Sea Lance, like SUBROC, is launched from the torpedo tubes of a submarine, rises above the surface of the water and flies to its target, whence it reenters the water and acts as a depth charge.

In addition, several new warhead concepts are in earlier phases of development, including:

- An earth penetrating warhead, which by exploding underground would improve the transfer of explosive energy to ground motion for destroying missile silos or hardened underground command centers;
- A warhead for a Maneuverable Reentry Vehicle (MaRV) to improve the accuracy of ballistic missile reentry vehicles and/or to evade possible future Soviet ballistic missile defenses; and
- A warhead for an improved version of the Lance short-range tactical missile.

## Soviet Nuclear Weapons Stockpile

### *Existing Weapons*

According to the Stockholm International Peace Research Institute (SIPRI), the Soviet Union has between 10,000 and 18,500 strategic nuclear warheads in its stockpile, with the higher figure assuming the existence of extra warheads as reloads for ICBM launchers and spares for submarine-launched missiles.[23] SIPRI credits the Soviet stockpile with another 9,400-13,700 warheads for theater nuclear forces and an additional unknown number of warheads for land-based surface-to-air missiles, artillery systems, torpedoes, and land-attack cruise missiles.

The 1987 SIPRI Yearbook lists 65 Soviet weapons systems that are capable of delivering nuclear warheads. The number of different warheads used by the Soviet Union is not known, but SIPRI lists 22 different yields or yield ranges for the warheads on the 65 delivery systems--from a low of less than 5 kt (kilotons) to a high of 1 Mt (megatons).[24] Tables 4 and 5 provide additional information about Soviet nuclear forces.

It is widely believed that Soviet warhead designs are not as advanced as U.S. designs.[25] For example, Soviet warheads are said to produce a smaller explosive yield for the same weight.[26] The Soviets make up for this deficiency by employing missiles with much larger carrying capacity than U.S. missiles, allowing them to deploy their higher yield weapons.

## Weapons Development

Little information is available about Soviet nuclear warheads under development. The Department of Defense publication Soviet Military Power lists several Soviet missile systems under development or beginning deployment. Some or all may include new warheads. They are listed below, along with DOD's March 1987 assessment of their state of development.[27]

- SS-N-23 -- New SLBM with more warheads and improved accuracy, compared to the SS-N-18 which it will replace; deployed in 1986;
- SSC-X-4 -- First Soviet ground-launched cruise missile, expected to become operational in 1987;
- SS-NX-21 -- Sea-launched cruise missile (SLCM), small enough to be fired from standard Soviet torpedo tubes, expected to become operational soon;
- SS-NX-24 -- Larger SLCM than the SS-NX-21, has been flight-tested, could become operational by 1988, a ground-based version may be developed;

- Replacement for SS-18 -- New ICBM with better accuracy and greater throw weight than the largest Soviet missile, has begun flight testing;
- Anticipated replacement for SS-24 -- New ICBM, perhaps larger, and with better accuracy and greater throw weight than the SS-24, the largest deployed Soviet solid-fueled missile, may begin flight-testing in the next few years;
- Anticipated MIRVed SS-25 -- A multiple-warhead version of the road-mobile missile (SS-25) currently being deployed. The new version could be developed later in this decade.

## U.S. Nuclear Weapons Testing Program

From 1945 through 1986 the United States announced 787 nuclear weapons tests.[28] Most U.S. tests are announced, but some low yield tests are not. The Natural Resources Defense Council (NRDC) estimates that the United States conducted an additional 117 unannounced tests through 1986.[29] In 1987, the United States conducted 15 announced tests.[30]

Since 1963, all but 10 U.S. tests have been conducted at the Nevada Test Site (NTS), an area larger than Rhode Island, located about 100 miles north of Las Vegas. The other 10 tests were conducted between 1963 and 1973 in Colorado, New Mexico, Mississippi, Nevada (outside of the NTS), and on Amchitka Island in Alaska.

The Department of Energy (DOE) is responsible for the research, development, testing and production of nuclear warheads in the United States. Its FY 1988 budget for those activities is $4.2 billion, comprised of $1.3 billion for research and development, $0.6 billion for testing and $2.2 billion for production. DOE's FY 1988 budget includes another $1.8 billion for production and processing of the special nuclear materials used in nuclear warheads, $0.9 billion to dispose of the waste from materials production activities, and $0.2 billion for security and safeguards.[31]

The Defense Nuclear Agency (DNA), an agency of the Department of Defense, performs research on the effects of

# Table 4

## Soviet Strategic Nuclear Forces, 1987

| Weapon system | | | | | Warheads | |
|---|---|---|---|---|---|---|
| Type | NATO code-name | No. deployed | Year deployed | Range (km) | Warhead × yield | No. in stockpile[a] |
| *ICBMs* | | | | | | |
| SS-11 Mod. 4 | Sego | 28 | 1966 | 11 000 | 1 × 1 Mt | 29 – 56 |
| Mod. 2 | | 360 | 1973 | 13 000 | 1 × 1 Mt | 380 – 720 |
| Mod. 3 | | 60 | 1973 | 10 600 | 3 × 250-350 kt (MRV) | 190 – 360 |
| SS-13 Mod. 2 | Savage | 60 | 1972 | 9 400 | 1 × 600-750 kt | 63 – 120 |
| SS-17 Mod. 2 | Spanker | 150 | 1979 | 10 000 | 4 × 750 kt (MIRV) | 630 – 1 200 |
| SS-18 Mod. 4 | Satan | 308 | 1979 | 11 000 | 10 × 550 kt (MIRV) | 3 200 – 6 200 |
| SS-19 Mod. 3 | Stiletto | 360 | 1979 | 10 000 | 6 × 550 kt (MIRV) | 2 300 – 4 300 |
| SS-X-24 | Scalpel | .. | 1987? | 10 000 | 7-10 × 100 kt (MIRV) | . . – . . |
| SS-25 | Sickle | 72 | 1985 | 10 500 | 1 × 550 kt | 76 – 140 |
| Total | | 1 398 | | | | 6 900 –13 000 |
| *SLBMs* | | | | | | |
| SS-N-5 | Sark | 39 | 1963 | 1 400 | 1 × 1 Mt | 41 – 47 |
| SS-N-6 Mod. 1/2 | Serb } | 288[b] | 1967 | 2 400 | 1 × 1 Mt } | 450 – 520 |
| Mod. 3 | | | 1973 | 3 000 | 2 × 200-350 kt (MRV) } | |
| SS-N-8 | Sawfly | 292 | 1973 | 7 800 | 1 × 800 kt-1 Mt | 310 – 350 |
| SS-N-17 | Snipe | 12 | 1977 | 3 900 | 1 × 1 Mt | 13 – 14 |
| SS-N-18 Mod. 1/3 | Stingray } | 224 | 1978 | 6 500 | 3-7 × 200-500 kt } | 710 – 1 900 |
| Mod. 2 | | | 1978 | 8 000 | 1 × 450 kt-1 Mt } | |
| SS-N-20[c] | Sturgeon | 80 | 1983 | 8 300 | 6-9 × 350-500 kt | 500 – 860 |
| SS-N-23[c] | Skiff | 32 | 1986 | 7 240 | 10 × 350-500 kt | 340 – 380 |
| Total | | 967 | | | | 2 400 – 4 100 |
| *Bombers* | | | | | | |
| Tu-95 | Bear A/B/C/G | 100 | 1956 | 8 300 | 2-4 × bombs/ASMs | 280 – 560 |
| Tu-95 | Bear H[d] | 40 | 1984 | 8 300 | 8 × AS-15 ALCMs | 320 – 640 |
| Total[e] | | 140 | | | | 600 – 1 200 |
| *Refuelling aircraft* | | | | | | |
| f | | 140-170 | | | | |
| *ABMs* | | | | | | |
| ABM-1B | Galosh Mod. | 32 | 1986 | 320 | 1 × unknown | 32 – 64 |
| ABM-3 | Gazelle | 68 | 1985 | 70 | 1 × low yield | 68 – 140 |
| Total | | 100 | | | | 100 – 200 |

[a] Figures for numbers of warheads are low and high estimates of possible force loadings (including reloads). Reloads for ICBMs are 5 per cent and 100 per cent; and for SLBMs 5 per cent and 20 per cent extra missiles and associated warheads. Half the SS-N-6s are assumed to be Mod. 3s, and SS-N-18 warheads are assumed to be 3 or 7 warheads. Bomber warheads are force loadings and force loadings plus 100 per cent reloads. It is assumed that 40 Bear Gs are now deployed (4 warheads each). All warhead total estimates have been rounded to two significant digits. Warhead estimates do not include downloading for single-warhead SS-17 Mod. 2, SS-19 Mod. 2 or SS-18 Mod. 1/3 missiles, which could be deployed, nor lower estimates for the SS-18 force, which could still include some Mod. 2 missiles with 8 or 10 warheads.

[b] It is not known whether the Soviet Union has already removed—or is planning to remove—from operational service an additional one or two Yankee Is during 1986 to make room for additional Typhoon and Delta IV Class submarines which may have entered sea trials. Alternatively, the USSR may have decided to wait to make these withdrawals until the USA exceeds the SALT limits.

[c] An additional Typhoon (20 SS-N-20 missiles) and Delta IV (16 SS-N-23 missiles) may be on sea trials and are thus included in the force totals. See note b.

[d] It is believed that, as of mid-1986, three squadrons of 12 Bear H aircraft each were in service. An additional squadron may have entered the operational force by the end of 1986.

[e] Excludes 30 MYA-4 Bison bombers which are under dispute. The USA believes that they remain SALT-accountable, while the USSR claims that they have been converted to refuelling tankers. Here they are included in the refuelling aircraft totals.

[f] Includes Badger and Bison A bombers converted to aerial refuelling and 15 confirmed new Bison conversions, with 30 possible new Bison conversions claimed by the USSR.

# Table 5

## Soviet Theater Nuclear Forces, 1987

| Weapon system | | No. deployed | Year deployed | Range (km) | Warheads Warhead × yield | No. in stockpile[a] |
|---|---|---|---|---|---|---|
| Type | NATO code-name | | | | | |
| **Land-based systems:** | | | | | | |
| *Aircraft* | | | | | | |
| Tu-26 | Backfire | 144 | 1974 | 3 700 | 2–3 × bombs or ASMs | 288 |
| Tu-16 | Badger | 287[b] | 1955 | 4 800 | 2 × bombs or ASMs | 480 |
| Tu-22 | Blinder | 136[b] | 1962 | 2 200 | 1 × bombs or ASMs | 136 |
| Tactical aircraft[c] | | 2 885 | .. | 700–1 000 | 1–2 × bombs | 2 885 |
| *Missiles* | | | | | | |
| SS-20 | Saber | 441 | 1977 | 5 000 | 3 × 250 kt | 1 323–2 200[d] |
| SS-4 | Sandal | 112 | 1959 | 2 000 | 1 × 1 Mt | 112 |
| SS-12 Mod. 1/2 | Scaleboard | ~130 | 1969/78 | 800–900 | 1 × 200 kt–1 Mt | 130 |
| SS-1C | Scud B | 690 | 1965 | 280 | 1 × 100–500 kt | 690–1 400 |
| SS-23 | Spider | | 1985 | 350 | 1 × 100 kt | |
| .. | FROG ? | | 1965 | 70 | 1 × 10–200 kt | |
| SS-21 | Scarab | 890 | 1978 | 120 | 1 × 20–100 kt | 890–3 600 |
| SS-C-1B[e] | .. | 100 | 1962 | 450 | 1 × 50–200 kt | 100 |
| SAMs[f] | | n.a. | 1956 | 40–300 | 1 × low kt | n.a. |
| *Other systems* | | | | | | |
| Artillery[g] | | <7 700 | 1974 | 10–30 | 1 × low kt | n.a. |
| ADMs | | n.a. | n.a. | – | n.a. | n.a. |
| **Naval systems:** | | | | | | |
| *Aircraft* | | | | | | |
| Tu-26 | Backfire | 132 | 1974 | 3 700 | 2–3 × bombs or ASMs | 264 |
| Tu-16 | Badger | 220 | 1961 | 4 800 | 1–2 × bombs or ASMs | 480 |
| Tu-22 | Blinder | 35 | 1962 | 2 200 | 1 × bombs | 35 |
| ASW aircraft[h] | | 204 | 1965 | .. | 1 × depth bombs | 204 |
| *Anti-ship cruise missiles* | | | | | | |
| SS-N-3 | Shaddock/Sepal | 264 | 1962 | 450 | 1 × 350 kt | 264 |
| SS-N-7 | .. | 96 | 1968 | 56 | 1 × 200 kt | 96 |
| SS-N-9 | Siren | 224 | 1969 | 111 | 1 × 200 kt | 224 |
| SS-N-12 | Sandbox | 120 | 1976 | 500 | 1 × 350 kt | 120 |
| SS-N-19 | .. | 112 | 1980 | 460 | 1 × 500 kt | 112 |
| SS-N-22 | .. | 44 | 1981 | 111 | 1 × 200 kt | 44 |

Source: SIPRI Yearbook 1987: World Armaments and Disarmament, p. 20-21. Oxford: Oxford University Press, 1987.

| | | | | | | |
|---|---|---|---|---|---|---|
| *Land-attack cruise missiles* | | | | | | |
| SS-N-21 | .. | ? | 1986 | 3 000 | 1 × n.a. | n.a. |
| SS-NX-24 | .. | 12? | 1986? | <3 000 | 1 × n.a. | n.a. |
| *ASW missiles and torpedoes* | | | | | | |
| SS-N-14 | Silex | 314· | 1968 | 50 | 1 × low kt | 314 |
| SS-N-15 | .. | n.a. | 1972 | 40 | 1 × 10 kt | n.a. |
| SUW-N-1/FRAS-1 | .. | 10 | 1967 | 30 | 1 × 5 kt | 10 |
| Torpedoes | .. | n.a. | 1957 | 16 | 1 × low kt | n.a. |
| *Naval SAMs*[f] | | | | | | |
| SA-N-1 | Goa | 65 | 1961 | 22–32 | 1 × 10 kt | 65 |
| SA-N-3 | Goblet | 43 | 1967 | 37–56 | 1 × 10 kt | 43 |
| SA-N-6 | .. | 33 | 1981 | 65 | 1 × 10 kt | 33 |
| SA-N-7 | .. | 9 | 1981 | 28–52 | 1 × 10 kt | 9 |

a    Estimates of total warheads are based on minimal loadings of delivery systems plus reloads for launchers which are deployed with reload weapons. Since many systems are dual-capable, these figures should not be viewed as precise. As a consequence, all figures (with exceptions for SS-20 and SS-4 missile force loading estimates since these systems only carry nuclear warheads) are rounded to two significant figures.

b    There are some 360 Badger strike variants, approximately two-thirds of which are Badgers.

c    Nuclear-capable tactical aircraft models include MiG-21 Fishbed, MiG-27 Flogger D/J, Su-7 Fitter A, Su-17 Fitter C/D/H, Su-24 Fencer and Su-25 Frogfoot.

d    The number of reload missiles available for each regiment is a matter of dispute. It is estimated that there is one missile reload available for two-thirds of the launchers in each regiment.

e    Land-based anti-ship missile.

f    Nuclear-capable land-based surface-to-air missiles probably include SA-1 Guild, SA-2 Guideline, SA-3 Goa, SA-5 Gammon, SA-10 Grumble and SA-12 Gladiator.

g    Artillery include some 3700 M-1981 2S5 152-mm SP guns, M-1976 152-mm T guns, M-1975 2S7* 203-mm SP guns and M-1975 2S4* 240-mm SP mortars. An additional 4000 M-1973 2S3 152-mm SP howitzers and older 152-mm towed guns may be nuclear-capable, although the status of crew certification for these systems is unknown. The 152-mm guns deployed on Sverdlov cruisers could also be nuclear-capable, although the status of the cruisers themselves is unclear.

h    Includes 94 Be-12 Mail, 50 Il-38 May and 60 Tu-142 Bear F. Land- and sea-based helicopters—including the Ka-25 Hormone, Ka-27 Helix and the Mi-14 Haze—could also have a nuclear delivery capability.

i    The SA-N-1, SA-N-3 and SA-N-6 are believed to have a definite nuclear capability and the SA-N-7 a possible nuclear capability. Number deployed is the number of launch arms (e.g., two twin launchers equal four launch arms) deployed on ships. Overall, there are more than 3300 SAMs of these four types deployed on 70 ships of 11 classes.

nuclear weapons. It conducts two to three nuclear weapons effects tests per year. DNA's FY 1988 budget is $311 million.[32]

Nuclear explosive tests are conducted for several purposes. DOE identified 606 of the 787 announced tests through 1986 as "weapons related."[33] These are generally to determine if new or modified warheads will work as they are intended, but include tests to examine new concepts in and improve understanding of nuclear weapons design. "Weapons effects" tests accounted for 89 of the remaining tests. These are mainly used to examine the effects of radiation from nuclear explosions on targets or equipment that may have to function during a nuclear war, including nuclear weapons and communications equipment. Of the remaining announced tests, 33 were to show that accidental detonation of the chemical high explosive in a warhead would not lead to significant nuclear yields; 27 were related to using nuclear explosives for large earth moving projects (Project Plowshare); 7 were to improve U.S. ability to detect nuclear explosions; and 19 were conducted for the United Kingdom.[34] Additional statistics on U.S. nuclear tests are shown in Table 6.

Nuclear warheads are explosively tested at several stages in their lifetime. Depending upon how radically a new warhead differs from previous designs, from one to several tests may be conducted during the design and development stages. Once a design is made final, the warhead is tested again before it enters production. Lastly, some time after a new warhead enters the active stockpile, a production verification test of one of the early units is held to ensure that production line models placed in the field work the same way as the specially fabricated and handled warheads tested during development. On average, a total of 4-6 tests are required before a new design is qualified to enter the stockpile.[35] Once a warhead has been placed in the stockpile, no further nuclear tests are routinely scheduled. However, additional nuclear tests may be conducted if nonnuclear reliability tests or examinations uncover a problem, if the design is modified, or if a nuclear test on a similar new warhead raises questions about the reliability of a stockpiled warhead.

It is useful to place nuclear explosive testing of warheads in context with other testing of nuclear weapons. Actual nuclear explosions represent only a few of the thousands of tests

## Table 6

## Announced United States Nuclear Tests

| TOTALS BY YEAR | | |
|---|---|---|
| PRE-TREATY | U.S. | US/UK |
| CY 1945 | 3 | |
| CY 1946 | 2 | |
| CY 1947 | | |
| CY 1948 | 3 | |
| CY 1949 | | |
| CY 1950 | | |
| CY 1951 | 16 | |
| CY 1952 | 10 | |
| CY 1953 | 11 | |
| CY 1954 | 6 | |
| CY 1955 | 18 | |
| CY 1956 | 18 | |
| CY 1957 | 32 | |
| CY 1958 | 77 | |
| CY 1959 | | |
| CY 1960 | | |
| CY 1961 | 10 | |
| CY 1962 | 96 | 2 |
| CY 1963 | 29 | |
| PRE-TREATY | 331 | 2 |
| POST-TREATY | | |
| CY 1963 | 14 | |
| CY 1964 | 29 | 1 |
| CY 1965 | 28 | 1 |
| CY 1966 | 40 | |
| CY 1967 | 28 | |
| CY 1968 | 33 | |
| CY 1969 | 29 | |
| CY 1970 | 30 | |
| CY 1971 | 12 | |
| CY 1972 | 8 | |
| CY 1973 | 9 | |
| CY 1974 | 7 | 1 |
| CY 1975 | 16 | |
| CY 1976 | 15 | 1 |
| CY 1977 | 12 | |
| CY 1978 | 12 | 2 |
| CY 1979 | 14 | 1 |
| CY 1980 | 14 | 3 |
| CY 1981 | 16 | 1 |
| CY 1982 | 18 | 1 |
| CY 1983 | 14 | 1 |
| CY 1984 | 12 | 2 |
| CY 1985 | 15 | 1 |
| CY 1986 | 12 | 1 |
| POST-TREATY | 437 | 17 |
| TOTAL | 768 | 19 |

| TOTALS BY TYPE | | |
|---|---|---|
| | U.S. | US/UK |
| TUNNEL | 58 | |
| SHAFT | 484 | 19 |
| CRATER | 9 | |
| TOTAL UG | 551 | 19 |
| AIRBURST | 1 | |
| BALLOON | 25 | |
| TOWER | 56 | |
| AIRDROP | 54 | |
| ROCKET | 12 | |
| SURFACE | 28 | |
| BARGE | 36 | |
| TOTAL ATMOS. | 212 | |
| TOTAL UW | 5 | |
| TOTAL | 768 | 19 |

| TOTALS BY DETECTION OF RADIOACTIVITY | |
|---|---|
| NO RADIOACTIVITY DETECTED | |
| NTS | 430 |
| DETECTION ON-SITE ONLY | |
| NTS | 93 |
| BOMBING RANGE | 1 |
| TOTAL | 94 |
| DETECTION OFF-SITE | |
| NTS | 136 |
| BOMBING RANGE | 4 |
| TOTAL | 140 |

| TOTALS BY LOCATION | | |
|---|---|---|
| | U.S. | US/UK |
| PACIFIC | 4 | |
| JOHNSTON ISL. AREA | 12 | |
| ENEWETAK | 43 | |
| BIKINI | 23 | |
| CHRISTMAS ISL. AREA | 24 | |
| TOTAL PACIFIC | 106 | |
| NTS UNDERGROUND | 540 | 19 |
| NTS ATMOSPHERIC | 100 | |
| TOTAL NTS | 640 | 19 |
| TOTAL S. ATLANTIC | 3 | |
| CENTRAL NEVADA | 1 | |
| AMCHITKA | 3 | |
| ALAMOGORDO | 1 | |
| JAPAN | 2 | |
| CARLSBAD | 1 | |
| HATTIESBURG | 2 | |
| FARMINGTON | 1 | |
| GRAND VALLEY | 1 | |
| RIFLE | 1 | |
| FALLON | 1 | |
| BOMBING RANGE | 5 | |
| TOTAL OTHER | 19 | |
| TOTAL | 768 | 19 |

| TOTALS BY PURPOSE | |
|---|---|
| | TOTAL |
| COMBAT | 2 |
| SAFETY EXPER. | 33 |
| STORAGE-TRANSP. | 4 |
| VELA UNIFORM | 7 |
| PLOWSHARE | 27 |
| WEAPONS RELATED | 606 |
| WEAPONS EFFECTS | 89 |
| JOINT US-UK | 19 |
| TOTAL | 787 |

Source: Office of Public Affairs, U.S. Department of Energy, Nevada Operations Office. Announced United States Nuclear Tests: July 1945 Through December 1986, p. 1. Jan. 1987. NVO-209 (rev. 7).

conducted annually. The components of a nuclear warhead, other than the nuclear package itself, are tested separately to ensure that they will function as intended. As but one example, the high explosive charge of the primary is detonated with the inner core(s) of uranium, plutonium, or both, modified or removed to verify its performance.[36] While the complete functioning of a nuclear warhead cannot be tested in a nonnuclear manner, nonnuclear tests and inspections can ensure quality control and verify that units are fabricated as designed. In addition, the integrated functioning of the warhead and its delivery system, except for the nuclear explosion, can be tested in a variety of conditions, including some that simulate elements of the operational environment.

A similar situation exists regarding nuclear weapons effects testing. Because of the expense in conducting nuclear tests, the resistance of military equipment to most nuclear weapons effects--e.g., blast, heat, and electromagnetic radiation--is tested through nonnuclear simulations. However, some effects, especially X-rays, cannot be fully simulated without nuclear explosions, and to ensure resistance against those effects, equipment is subjected to nuclear tests.

Weapons design and physics tests are conducted in vertical shafts drilled deep into the ground. The weapon to be tested, along with instrumentation to monitor the test, is placed in a shaft some 600 feet to 3,000 feet deep. The shaft is filled with gravel, sand, and several epoxy plugs that are impervious to gas. After the epoxy has hardened, the weapon is detonated. An average weapons development test costs about $20 to $30 million.[37]

Weapons effects tests are more complicated, require more extensive tunnelling, and cost $40 to $70 million per test.[38] Most effects tests take place in horizontal tunnels drilled about 1,000 feet into a solid rock mesa. For the most common type of effects test a small nuclear device is placed on a platform at the far end of the tunnel in front of a large steel cone that runs the length of the tunnel. The steel cone is narrow near the nuclear device and increases to several feet in diameter near the entrance to the tunnel. The equipment to be tested is put in several positions near the wide end of the cone. Following the detonation of the nuclear device, large doors located near the explosion are driven shut by the force of the explosion, letting the radiation through to the equipment beyond, but preventing the release of radioactive gases and debris.[39]

DOE announces the yields of U.S. nuclear tests as either being below 20 kt or between 20 kt and 150 kt. However, a graph showing an estimate of the relative yield of all U.S. tests from 1980 through 1984 was published in the Journal of the Federation of American Scientists and is reproduced in Figure 2. The most interesting feature of the graph is that about half of the tests conducted had yields between 5 kt and 20 kt (shaded area), which has been interpreted to indicate that the yield of fission primaries for thermonuclear devices falls in that range.[40]

## Soviet Nuclear Weapons Testing Program

The Natural Resources Defense Council (NRDC) estimates that from 1949 through 1985 the Soviet Union conducted 608 nuclear explosions.[41] Prior to August 5, 1963, when the Limited Test Ban Treaty was signed, all but three of the Soviet Union's 184 nuclear explosions were in the atmosphere. Since then, all have been underground.

The main Soviet weapons test site is near the city of Semipalatinsk in eastern Kazakhstan. Since 1963, 32 tests (an average of less than 3 per year) have been performed on the island of Novaya Zemlya, above the Arctic Circle. In addition, 117 of the 424 post-1963 Soviet nuclear explosions listed by NRDC were conducted at locations outside those two tests sites--mostly in west Kazakhstan, in the region of the Ural Mountains, and in Siberia. The Soviet Union has announced civil engineering missions for some of those explosions, such as assisting in earth-moving, creating underground reservoirs, and stimulating the production of oil and gas. NRDC considers 102 of the 117 nuclear explosions conducted outside the test sites since 1963 as having been for peaceful purposes.[42] However, nuclear explosions for any purpose can produce useful weapons information. The United States has not conducted any civil nuclear explosions since

**Figure 2**

**Distribution of Explosive Yields at NTS:
1980 Through 1984**

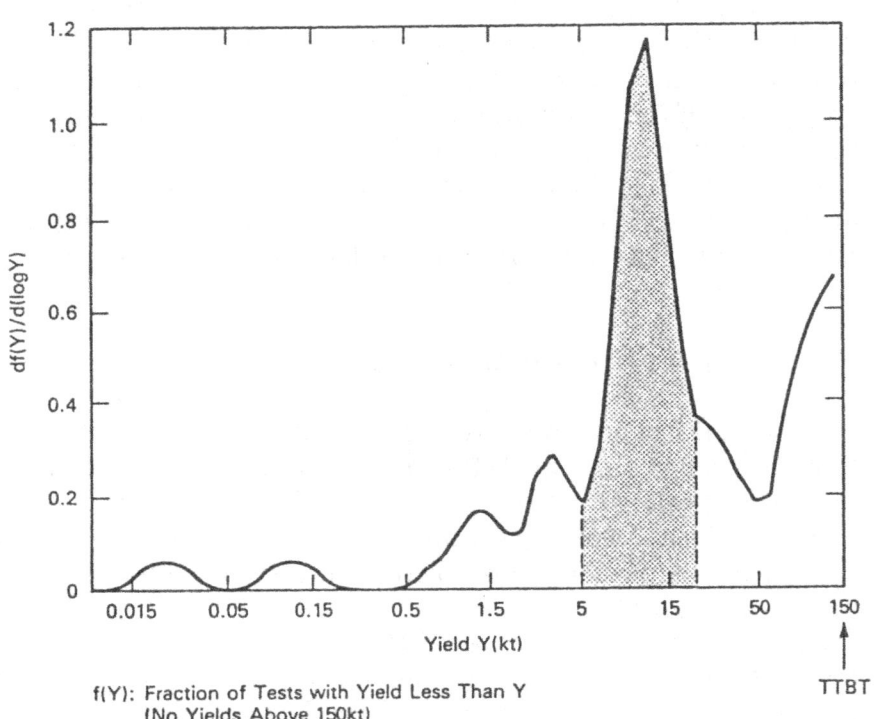

f(Y): Fraction of Tests with Yield Less Than Y
(No Yields Above 150kt)

TTBT

This curve shows the relative frequency of nuclear tests at each yield for all tests at the Nevada Test Site (NTS) from 1980 through 1984. The vertical scale produces an area under the curve of one so that the relative frequency of a test during this period having a given yield can be seen immediately.

Source: Ray E. Kidder. "Militarily Significant Nuclear Explosive Yields." Journal of the Federation of American Scientists, v. 38, Sept. 1985: 3. Reproduced with the permission of the Federation of American Scientists. Copyright 1985. All rights reserved.

## Table 7

## Summary of Known Soviet Nuclear Explosions, 1949-1985

| Yr. | No. | PRESUMED PURPOSE Military | Peaceful | LOCATION 1 | 2 | 3 | Cumulative Total | Announced by USG |
|---|---|---|---|---|---|---|---|---|
| 1949 | 1 | 1 | 0 | 0 | 0 | 1 | 1 | 1 |
| 1950 | 0 | 0 | 0 | 0 | 0 | 0 | 0 | 0 |
| 1951 | 2 | 2 | 0 | 2 | 0 | 0 | 3 | 2 |
| 1952 | 0 | 0 | 0 | 0 | 0 | 0 | 3 | 0 |
| 1953 | 4 | 4 | 0 | 4 | 0 | 0 | 7 | 2 |
| 1954 | 7 | 7 | 0 | 0 | 0 | 7 | 14 | 1 |
| 1955 | 5 | 5 | 0 | 2 | 0 | 3 | 19 | 4 |
| 1956 | 9 | 9 | 0 | 0 | 0 | 9 | 28 | 7 |
| 1957 | 15 | 15 | 0 | 2 | 4 | 9 | 43 | 13 |
| 1958 | 29 | 29 | 0 | 0 | 26 | 3 | 90 | 25 |
| 1959 | 0 | 0 | 0 | 0 | 0 | 0 | 90 | 0 |
| 1960 | 0 | 0 | 0 | 0 | 0 | 0 | 90 | 0 |
| 1961 | 50 | 50 | 0 | 6 | 24 | 20 | 140 | 50 |
| 1962 | 44 | 44 | 0 | 10 | 32 | 2 | 184 | 40 |
| 1963 | 0 | 0 | 0 | 0 | 0 | 0 | 184 | 0 |
| 1964 | 6 | 6 | 0 | 4 | 2 | 0 | 190 | 3 |
| 1965 | 9 | 8 | 1 | 8 | 0 | 1 | 199 | 4 |
| 1966 | 15 | 13 | 2 | 12 | 1 | 2 | 214 | 7 |
| 1967 | 17 | 16 | 1 | 15 | 1 | 1 | 231 | 4 |
| 1968 | 13 | 11 | 2 | 12 | 1 | 2 | 244 | 7 |
| 1969 | 16 | 12 | 4 | 10 | 1 | 5 | 260 | 12 |
| 1970 | 17 | 15 | 2 | 9 | 1 | 7 | 277 | 10 |
| 1971 | 19 | 15 | 4 | 11 | 1 | 7 | 296 | 14 |
| 1972 | 22 | 14 | 8 | 12 | 1 | 9 | 318 | 14 |
| 1973 | 14 | 10 | 4 | 6 | 3 | 5 | 332 | 14 |
| 1974 | 19 | 16 | 3 | 13 | 3 | 3 | 351 | 8 |
| 1975 | 15 | 13 | 2 | 10 | 3 | 2 | 366 | 10 |
| 1976 | 17 | 15 | 2 | 13 | 2 | 2 | 383 | 10 |
| 1977 | 18 | 15 | 3 | 11 | 2 | 5 | 413 | 11 |
| 1978 | 28 | 21 | 7 | 18 | 2 | 8 | 441 | 20 |
| 1979 | 29 | 21 | 8 | 19 | 2 | 8 | 470 | 15 |
| 1980 | 21 | 18 | 3 | 17 | 1 | 3 | 491 | 10 |
| 1981 | 22 | 16 | 5 | 16 | 1 | 5 | 513 | 9 |
| 1982 | 31 | 15 | 16 | 14 | 1 | 16 | 544 | 6 |
| 1983 | 27 | 14 | 13 | 12 | 2 | 13 | 571 | 9 |
| 1984 | 28 | 18 | 10 | 16 | 1 | 11 | 599 | 17 |
| 1985 | 9 | 7 | 2 | 7 | 0 | 2 | 608 | 4 |
| Totals | 608[a] | 506[a] | 102 | 289 | 118 | 171[+] | 608[a] | 363 |

a These figures include 18 tests conducted 1956-1958, and 12 conducted 1963-1977, for which the year is not known; all presumed military.

1: Semipalatinsk; 2: Novaya Zemlya; 3: Other; USG: U.S. Government.

Source: Sands, J.I., R.S. Norris, and T.B. Cochran. Known Soviet Nuclear Explosions, 1949-1985. Revised Preliminary List. Nuclear Weapons Databook Working Papers. NWD 86-3. Washington, Natural Resources Defense Council. February 1986, revised June 1986. Table 2. Used by permission.

Figure 3

**Yield Distributions for Soviet Nuclear Explosions**

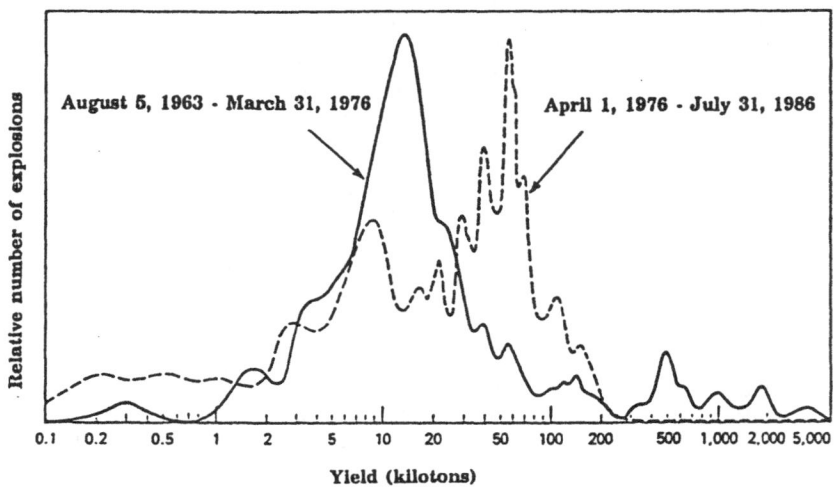

Note: While the measurement of yields of all nuclear explosions is subject to uncertainty, this uncertainty is especially large for yields below approximately 5 kt.

Source: Lynn R. Sykes and Dan M. Davis. "The Yields of Soviet Strategic Weapons." Scientific American, v. 256, Jan. 1987: 33. Copyright 1986 by Scientific American, Inc. All rights reserved.

1973. Additional information on Soviet nuclear explosions is given in Table 7.

On August 6, 1985, the Soviet Union declared a unilateral moratorium on nuclear weapons testing.[43] Late in 1986, Soviet officials stated that the moratorium would end if the United States continued testing, and on February 26, 1987, following two U.S. tests that month, the Soviets announced they had conducted a nuclear test earlier that day.

There has been considerable controversy over the yields of Soviet nuclear explosions, particularly with regard to whether explosions since March 31, 1976, when the 150-kt limit of the Threshold Test Ban Treaty (TTBT) and the Peaceful Nuclear Explosions Treaty (PNET) took effect, have violated that limit. That question is addressed in detail in Chapter 7. Setting that question aside, several conclusions relevant to this section can be drawn from Figure 3, a comparison of the yields of Soviet nuclear explosions before and after March 31, 1976:

- Soviet explosions with yields <u>well above</u> 150 kt were conducted prior to 1976, but apparently ceased at that date;
- The Soviet Union, like the United States, conducts a large number of explosions with yields between 5 kt and 20 kt, some of which are presumed to be tests of primaries for thermonuclear warheads; and
- The relative number of Soviet explosions in the 40 kt-150 kt range increased markedly after 1976, which can be interpreted as an increase in full yield testing of much smaller warheads than relied upon previously and/or partial yield testing of warheads with full yield above 150 kt.

## Notes

1. Unless cited to another source, all information in this section on the design and functioning of nuclear weapons comes from Morland, Howard. The H-Bomb Secret. The Progressive, vol. 43, Nov. 1979: 14-23; and DeVolpi, Alexander, et al. Born Secret: The H-Bomb, the Progressive Case and National Security. Pergamon, New York, 1981.

2. U.S. Department of Defense and Department of Energy. The Effects of Nuclear Weapons. Samuel Glasstone and Philip Dolan, ed. Washington, Govt. Print. Off., 1977. p. 17.

3. Glasstone and Dolan, The Effects of Nuclear Weapons, p. 17.

4. Morland, The H-Bomb Secret, p. 14-23.

5. Cochran, Thomas B., et al. U.S. Nuclear Forces and Capabilities. Nuclear Weapons Databook, vol. 1. Cambridge, MA, Ballinger, 1984. p. 31. (Hereafter cited as Nuclear Weapons Databook, v. 1.)

6. Despite this problem, a fission device with a yield of 500 kt has been detonated by the United States. Nuclear Weapons Databook, v. 1, p. 34.

7. Morland, The H-Bomb Secret, p. 14-23.

8. Soviet weapons must contain fuzing and firing mechanisms, and may contain the other components, but detailed design information about Soviet weapons is unavailable.

9. Stockholm International Peace Research Institute. World Armaments and Disarmament: SIPRI Yearbook 1987. Oxford University Press, New York, 1987. Tables 1.1 and 1.2. (Hereafter cited as SIPRI Yearbook 1987.) p. 6-7.

10. Cochran, Thomas B., et al. Nuclear Weapons Databook. v. 2. U.S. Nuclear Warhead Production. Cambridge, MA, Ballinger, 1987. Table 1.2. (Hereafter cited as Nuclear Weapons Databook, v. 2.) p. 10-11.

11. U.S. Nuclear Weapons Stockpile (June 1987). Table printed in the Bulletin of the Atomic Scientists, June 1987. (Reprinted here as Table 1). p. 56.

12. U.S. Congress. House. Committee on Appropriations. Energy and Water Development Appropriations for 1988. Hearings, 100th Cong., 1st Sess., Part 6. Washington, Govt. Print. Off., 1987. p. 553. (Hereafter cited as FY88 House Energy and Water Appropriation Hearing.)

13. Lynch, David J. U.S. Takes Wraps Off Huge Bomb. Defense Week, vol. 8, no. 31, Aug. 3, 1987: 1.

14. FY88 House Energy and Water Appropriations Hearings. p. 823.

15. Nuclear Weapons Databook, v. 1, p. 36.

16. According to John Immele, Deputy Associate Director for Nuclear Design, Lawrence Livermore National Laboratory, the IHE B61 mods are an exception to this rule as their development was motivated by the need for improved safety in the Air Force's main multipurpose bomb. Telephone conversation with Robert Civiak.

17. Barker, Robert B. Deputy Assistant Director. Arms Control and Disarmament Agency. Briefing to congressional staff. August 28, 1986.

18. Nuclear Weapon R & D and the Role of Nuclear Testing. Energy and Technology Review, Lawrence Livermore National Laboratory, September 1986: 9.

19. Fetter, Steve. Toward a Comprehensive Test Ban. Ballinger, Cambridge, MA. (in press).

20. Phases 2a and 7 were added later. Department of Energy Congressional Budget Request, FY87. January 1987, 4 vol. DOE/MA-0274 v. 1, p. 19.

21. FY88 House Energy and Water Appropriation Hearing. p. 796, 823.

22. FY88 House Energy and Water Appropriation Hearing. p. 829.

23. SIPRI Yearbook, 1987, Table 1.4.

24. SIPRI Yearbook, 1987. p. 18-21.

25. SIPRI Yearbook, 1986. p. 13.

26. Sykes and Davis, The Yields of Soviet Strategic Weapons, p. 29-37.

27. U.S. Department of Defense. Soviet Military Power, 1987. 6th ed. Washington, Govt. Print. Off., 1987. 159 p. Most data is from this source, p. 19-38. Deployment year of the SS-N-23 is from Collins, John, and Bernard Victory, U.S./Soviet Military Balance: Statistical Trends, 1977-1986 (As of January 1, 1987). U.S. Library of Congress. Congressional Research Service. Report 87-745 S. Washington, Sept. 1987, p. 19.

28. U.S. Department of Energy. Nevada Operations Office. Announced United States Nuclear Tests. July 1945 through December 1986. NVO-209 (rev. 7). January 1987. 66 p.

29. Norris, R.S., T.B. Cochran, and W.M. Arkin. Known U.S. Nuclear Tests, July 1945 to 31 December 1987. Nuclear Weapons Databook Working Papers. NWD 86-2 (Rev. 2A). Washington, D.C., Natural Resources Defense Council. January 1988. 60 p.

30. Ibid.

31. U.S. Department of Energy. Congressional Budget Request, FY 1989. Volume 1, Atomic Energy Defense Activities. February 1988.

32. U.S. Congress. House. Making Further Continuing Appropriations for the Fiscal Year Ending September 30, 1988. Conference Report to Accompany H.J. Res. 395. H. Rept. 100-498. 100th Congress, 1st Session. Washington, Govt. Print. Off., 1987. p. 656.

33. Announced United States Nuclear Tests, p. 1.

34. The United States currently conducts all of the United Kingdom's nuclear explosive tests.

35. This figure can be gleaned from the facts that the United States has conducted 606 weapons related tests, some of which were not related to specific warheads and some to warheads already in the stockpile, and that the United States has designed about 100 unique warheads.

36. Morland, The H-Bomb Secret, p. 22.

37. Norris, R.S., et al. Known U.S. Nuclear Tests, July 1945 to 31 December 1987. op. cit. p. 15.

38. Ibid.

39. Horgan, John. Underground Nuclear Weapons Testing. IEEE Spectrum, vol. 23, April 1986: 32-43.

40. Feiveson, Harold A., et al. A Low-Threshold Test Ban Treaty. Science, vol. 238, Oct. 23, 1987: 455-459.

41. Sands, J.I., R.S. Norris, and T.B. Cochran. Known Soviet Nuclear Explosions, 1949-1985. Revised Preliminary List. Nuclear Weapons Databook Working Papers. NWD 86-3. Washington, D.C., Natural Resources Defense Council. February 1986, revised June 2, 1986. 50 p. The 1987 SIPRI Yearbook lists only 597 Soviet nuclear explosions over that period.

42. Sands, J.I., et al. Known Soviet Nuclear Explosions, 1949-1985. p. 44.

43. The last Soviet nuclear explosion prior to the moratorium took place on July 25, 1985.

# 4

## Effects of More Restrictive Nuclear Test Bans on New Weapon Development

*David Cheney, Robert Civiak,*
*Jonathan Medalia, and Paul Zinsmeister*

### Introduction

This chapter discusses how a more restrictive ban on nuclear testing--either a comprehensive test ban treaty (CTBT), a low-yield threshold treaty (LYTT), or a quota test ban treaty (Quota)--might affect the development of current and future generation nuclear weapons and the potential consequences for strategic stability. Whether the effect of a more restrictive ban on nuclear testing on new nuclear weapons is desirable is the paramount issue of the test ban debate. Those who support more severe restrictions on testing hope to stop the competition between the United States and the Soviet Union in deploying improved nuclear weapons and, in so doing, reduce the risk of war between the two sides. On the other side are those who believe it essential for deterrence that the United States continue to exploit its technological advantage and deploy improved nuclear weapons that counter Soviet threats to the survivability and capability of U.S. nuclear forces. Nuclear testing, they argue, is absolutely essential to this pursuit.

In general, a CTBT would: (1) preclude deployment of nuclear warheads that are significantly different from those currently deployed or in advanced stages of development; (2) curtail modifications to warheads; (3) constrain deployment of new delivery systems by impeding warhead design; and (4) prevent deployment of future generation nuclear weapons. The effects of a LYTT or a Quota would depend on its terms, but in general would be less severe than a CTBT.

## Technical Background

This section discusses the uses of nuclear and nonnuclear testing in developing new current generation warheads and warheads or other nuclear devices for future generation weapons. Current generation warheads are those for which no special effort is made to modify the form or direction of the explosive energy release. Future generation weapons either enhance or suppress particular effects of nuclear explosions or use nuclear explosions as their energy source.

### Current Generation Nuclear Warheads

Nuclear testing is critical to the design and development of nuclear warheads because it provides detailed information on the performance of candidate designs that cannot be obtained in any other way. Warhead designers use this information in modifying a warhead design to optimize its performance and achieve the best possible match to the military requirements. Nuclear explosive testing also provides valuable assurance that warheads of the final design will perform as intended.

Numerous nonnuclear testing and simulation techniques are also vital in the design and evaluation of candidate warheads. Complete warheads minus their nuclear materials are tested. Flash X-ray equipment is used to take high speed X-ray snapshots of nonnuclear implosions within the warheads.[1] Many warhead components, such as fuzes and arming systems, are tested separately. Nonnuclear simulations of nuclear effects (see chapter 6) assist in designing warheads to be survivable.

Inertial confinement fusion--tiny fusion explosions initiated by heating and compressing a pellet of fuel with lasers or particle beams--is used to simulate some aspects of nuclear explosions. Most importantly, designers use high-speed supercomputers to run numerical simulations that predict the performance of warhead designs much more quickly and cheaply than can be done with nuclear tests, allowing numerous design variants to be "tested."[2] Computer simulations have improved in recent years as computers, knowledge of the processes in nuclear explosions, and calculation techniques have improved. Computer models are adequate to give warhead designers confidence in the performance of warheads that have been altered in minor ways from nuclear-tested models, but cannot now provide the same degree of confidence in totally new warheads as nuclear testing.

Nonnuclear and, sometimes, nuclear tests are also used in the many instances when an existing warhead is to be modified. Modifications are made for many reasons: a defect that emerges with age may need to be fixed; the armed services may want to use an existing warhead in a different way, e.g., use a warhead from an antiaircraft missile as an antisubmarine warfare weapon; a different chemical explosive may be needed to avert a safety hazard; changed deployment plans for a warhead may make a new permissive action link (PAL) desirable; or a warhead may have to be used under differing conditions than those for which it was designed. While nuclear tests are needed to design new warheads, they may or may not be needed for modifications to existing warheads.

The reason for discussing modifications in a chapter on the impact of a more restrictive test ban on new warheads is twofold: some modifications require nuclear testing, and the distinctions between modifications and new warheads can be fuzzy. Regarding the first point, the line dividing what a CTBT precludes from what it permits runs through modifications. While a CTBT would preclude deployment of any totally new warheads and some modifications, other modifications could be carried out under a CTBT. Thus, to understand the limits of a CTBT's effects on warhead development, one must understand what kinds of modifications can and cannot be made without nuclear testing.

In general, changes that do not affect the physics package (explosive component) of the warhead can confidently be made without nuclear testing. An example is the introduction of certain types of PALs that are outside the physics package. Some types of substitutions that would change the yield of a warhead can also be made without testing, such as changes in the isotope ratio of the fuel in the secondary. Other changes to the physics package of warheads that are within the range of designers' experience and computer modeling capability might also be made.

Many substantial changes cannot confidently be made without testing. For example, developing new primaries that use new high explosives or new permissive action links that are an integral component of the warhead's physics package would require testing.[3] Other modifications that require significant changes to the physics package of weapons, such as modifying warheads to fit in very narrow reentry vehicles or to resist extraordinary environmental forces (such as those earth penetrating warheads would face), would generally require testing.

### Future Generation Nuclear Weapons

The types of future generation nuclear weapons receiving the most attention at present are those constituting the nuclear portion of the Strategic Defense Initiative (SDI). SDI is a research program focused on advanced technologies to defend against ballistic missiles. The Reagan Administration repeatedly stated its preference for defensive systems that do not use nuclear explosive devices either directly as weapons, or indirectly as power sources for new weapons. Nevertheless, SDI is exploring new weapons concepts that require nuclear explosions. The potential advantage of nuclear versions is that they are cheaper, smaller, and lighter, making them attractive for basing in space or for rapid launch into space in a crisis. The Administration maintained, however, that it examined nuclear weapons under the SDI program only: (1) to understand how the Soviet Union might employ nuclear weapons against a U.S. ballistic missile defense; (2) to understand how the Soviets might use nuclear weapons as a component of their own ballistic missile defense; and (3) to explore nuclear directed energy options as SDI possibilities if needed.[4]

The major SDI interest in nuclear explosives is to explore nuclear directed energy weapons (NDEW). NDEWs are nuclear weapons which, by directing a fraction of their energy into a relatively narrow beam, can potentially destroy missiles and warheads thousands of kilometers away. NDEW concepts currently being explored under SDI include X-ray lasers, focused microwaves, particle beams, optical lasers, and nuclear-driven hypervelocity pellet guns (or so-called nuclear shotguns). All of these weapons except the X-ray laser have nonnuclear counterparts. Nuclear and nonnuclear SDI components might complement each other: a sudden burst of energy from a nuclear directed energy weapon is best suited for attacking many targets at once, such as a Soviet salvo attack, while nonnuclear weapons are better suited for attacking a sequentially launched attack, such as might come from missile submarines.[5]

In addition to the future generation weapons under consideration for SDI, many other types of future generation nuclear weapons are conceivable, according to former weapons designer Theodore Taylor.[6] Weapons that enhance, suppress, or focus effects of nuclear weapons might be developed. In the same way that neutron bombs enhance the output of neutrons and minimize blast effects to reduce collateral damage, weapons might be designed to enhance gamma rays, blast, X-rays, visible light, infrared radiation, microwaves, or electromagnetic pulse. Just as an X-ray laser might focus X-rays into a narrow beam, other forms of electromagnetic radiation might also be channeled. Taylor writes that the military potential of focused microwave weapons is "awesome" because military electronic systems are highly sensitive to microwave radiation and because microwave radiation readily penetrates the atmosphere.[7] Researchers at Lawrence Livermore National Laboratory are conducting research relevant to microwave and other electromagnetic weapons.[8] It is not clear which, if any, novel types of weapons are possible or militarily useful. For example, it is difficult to determine the effectiveness of microwave weapons without atmospheric and space testing.

## The Effects of a CTBT

### Effects on the Development and Deployment of New Current Generation Warheads

*General Comments*. Supporters and opponents of continued testing agree that a comprehensive test ban would virtually stop the deployment of new current generation warheads because without testing warhead designers could not achieve the level of confidence currently required of new warheads.

While deployment of new warheads virtually would be stopped under a CTBT, existing designs could be adapted for new uses without nuclear testing.[9] Such adaptation may require minor modifications to the warhead design. Deciding which changes are acceptable would have to be made on a case-by-case basis, balancing the need for the upgraded performance of a new weapon against the degree of uncertainty introduced by modifying a warhead design without testing.

The safe approach is to make no modifications because even seemingly minor ones can cause warheads to fail.[10] If changes were made during a test ban, confidence in the new warhead would be lower than that currently achieved because a production verification test--a test of a production version of the warhead--could not be conducted. During the 1958-1961 testing moratorium, at least one modified warhead, the W52 for the Army's Sergeant surface-to-surface missile, was introduced into the stockpile without nuclear testing and failed when tested after the moratorium.[11] Several other warheads introduced into the stockpile with incomplete testing also developed problems.[12] As a result of such experience, current DOE practice is to conduct a production verification test after a warhead has been in its operating environment for a year or two.[13] Nonetheless, testing does not assure perfection. Warhead designers were surprised by one recent test in which a previously-certified warhead was tested at the low end of its design temperature range and produced only a small fraction of its expected yield.[14]

Nonetheless, there would probably be substantial pressure during an extended CTBT to make at least some minor modifications to existing warheads, such as to fix problems, save money, or make warheads fit better with new delivery vehicles. Decisionmakers would have to choose whether the benefits of a new warhead are worth the reduced confidence. There would be little point in using an untested warhead where a well-tested one could do almost as well with higher confidence. On the other hand, if a new warhead provided great benefits and seemingly low technical risk, it is possible that decisionmakers would accept reduced confidence to deploy it.

Making modifications without testing involves risks, which may be minimized but not eliminated with expert judgment. Yet the skill of designers to make the necessary judgments would likely decline under a CTBT because they could not hone their skills by nuclear testing. The risks of modifying warhead designs would grow if, as supporters of continued testing argue, a CTBT caused many experienced designers to leave weapons laboratories. At the same time, the pressure upon weapons designers to modify warheads to match new delivery systems might grow.

Given these limits, what weapons development work could proceed under a CTBT? Both the United States and the Soviet Union could conduct extensive research and development work on new advanced warheads, employing all means of nonnuclear testing and, if permitted, very low-yield nuclear testing. Even though such work would not under foreseeable conditions lead to deployment of new warheads during a CTBT, it could serve as a hedge and a deterrent against a breakout from the treaty by the other side.

A CTBT might permit testing at very low yields that are well below the limits of detection--for example, explosions in reusable indoor facilities or with yields of a few pounds of TNT. It may be possible to construct a contained, reusable, seismically quiet facility for conducting tests with yields up to about 0.3 kilotons.[15] The smallest warhead in the U.S. stockpile has a subkiloton yield; new warheads in that yield range could be tested inside such a facility. The military value of new warheads with yields below 0.3 kt is unclear, however. At even lower yields, hydronuclear experiments, or so-called zero-yield testing,[16] can provide confidence that nuclear fission explosions will work, but such tests cannot be used to test boosted fission warheads and thus may not aid militarily useful warhead development.

The greatest potential for developing new weapon systems under a CTBT would come from designing new delivery vehicles. While U.S. practice is to design a new warhead for a new weapon, delivery systems can be designed around existing warheads. As described in Chapter 3, the current U.S. stockpile includes 26 distinct warhead designs of widely varying size, weight, and yield, which are used with at least 37 delivery systems. In designing any engineering system, the goal is to obtain the optimum characteristics given the constraints. Designers always have constraints in available materials, costs, components available, and ultimately, physical principles. Under a test ban, they would have the additional constraint of making only minor modifications to available warheads. In some cases, to meet the added requirements, a new missile with an old warhead might have to be larger or more accurate than with a new warhead, adding to cost and making for a more complicated design. In other cases, an existing or slightly modified warhead would be adequate. For example, current plans are for the Midgetman missile to use the MX (W87) warhead, with minor modifications.[17]

An analogy to summarize the impact of a CTBT is the design of automobiles, with the engine analogous to the warhead and the rest of the automobile analogous to the delivery system. One can design a new engine to fit with a new automobile, or design a new automobile around an existing engine. The latter is often done, but constrains design and performance. If today's automobiles were designed around 1950s-era engines, many advances could have been made, but automobiles would likely be heavier, use more fuel, and pollute more than current versions.

*Specific Weapon Systems Under Development.* This section examines the effects of a CTBT on the prospects for deployment of several current generation nuclear weapon systems under development in the United States.

*The D-5 missile for the Trident submarine* and its warhead, the W88, are both in the final stages of development.[18] The W88 entered phase 4, production engineering, in April 1986, and its first production unit (phase 5) is scheduled for October 1988.[19] At this stage of development, the design of a warhead generally has been established and tested successfully. Under normal circumstances the only remaining nuclear test of the W88 would be a production verification test of an early production unit. A CTBT that took effect a few months after October 1988 or one that made provision for the final test of the W88 would not affect certification and deployment of the W88, as long as the current schedule could be met and the final test did not uncover a major unexpected problem. If a CTBT were to preclude the last scheduled test of the W88 it is likely that the warhead would be deployed anyway. However, confidence in its performance would not be as high as it would be if it had received the final test. In that case, the Navy might revisit the decision to place Trident II missiles on all its submarines: it might deploy some Trident IIs with W76 warheads, the ones used on the Trident I missile, or it might retain some Trident Is with W76 warheads. In addition to tests of the W88, underground nuclear effects tests are scheduled for September 1989 to verify the survivability and hardness of the missile and its electronics.[20] The consequences of precluding those tests are discussed in Chapter 6.

*The W82 Warhead for the 155mm Artillery Fired Atomic Projectile (AFAP)* entered phase 4 development in 1986.[21] Tests have shown that the W82, including its nonnuclear components, will survive the gun-firing environment of the 155mm howitzer for which it is designed.[22] It is likely that the only remaining test of the W82 is the final production verification test. Since the AFAP is a short-range weapon, it does not need to be hardened against X-rays and therefore no nuclear effects tests are scheduled for it. The future of this weapon system is unclear, however, even without further limits on nuclear testing. It was originally intended as an enhanced radiation weapon (i.e., neutron bomb),[23] but Congress prohibited production of that design. Congressional support for the improved, nonenhanced radiation version of the 155mm AFAP has been mixed, and no funds were requested or provided for production of the W82 in FY 1988. If it should be determined that the improved 155mm AFAP is needed, it is likely that a CTBT would not preclude its use, albeit with less than full confidence because of the lack of a final production verification test. Alternatively, a final test could be conducted before a CTBT entered into force.

*The Small ICBM (Midgetman)* is intended to improve the survivability of the United States' land-based strategic missile deterrent by virtue of its mobile basing. In December 1986, President Reagan authorized proceeding with full-scale development of the small ICBM (SICBM). The warhead for the SICBM was scheduled to enter full scale development (phase 3) in mid-1987. The President also stated that initial operating capability for the SICBM would be achieved in 1992.[24] However, as of April 1988, DOD had not authorized proceeding to phase 3 and was proposing cancellation of the SICBM program. According to testimony of James Culpepper, Acting DOE Assistant Secretary for Military Applications, a modified version of the W87 warhead (used on the MX missile) has been selected as the warhead for the SICBM.[25] Robert Barker stated at the same hearing that the use control features (i.e., advanced PAL) will be upgraded to be appropriate for a mobile missile.[26]
In addition, warhead yield will be greater than that of the W87 for MX.[27] The increased yield will not require additional developmental nuclear testing nor will the PAL upgrade as it is envisioned presently. Current practice would require a nuclear test for production verification if the W87 production line were shut down for more than a few months between production of the last W87 for MX (whether silo-based or rail garrison-based) and the first W87 for SICBM. This test would

insure that the production process has not been altered in such a way as to introduce flaws as a result of the shutdown.

*The Short Range Attack Missile (SRAM-II)* is an improved air-to-surface missile with greater range, smaller size, and improved safety. One improvement in the warhead, which would replace the W69 used in the SRAM-I, is the use of accident-resistant, insensitive high explosive (IHE).[28] The SRAM-II warhead entered phase 3 development in late 1987, and is scheduled to enter into production in July 1991.[29] Congress denied DOE's request for funding for the SRAM-II warhead in FY 1987, but SRAM-II funding has been approved for FY88. Since the SRAM-II warhead has just begun phase 3 development and at least one change (the addition of IHE) is a major change, it is unlikely that development could be completed with sufficient confidence for deployment of the new warhead under an early CTBT. It is possible, however, that a new missile could be developed using the W69 or some other warhead already in the stockpile. In that case, the safety of IHE might be sacrificed and the weapon also might not be as capable as a SRAM-II with a new warhead.

*The Antisubmarine Warfare Nuclear Depth/Strike Bomb (ASW ND/SB)* is intended to be a dual-capable bomb delivered by aircraft for use as a depth bomb against submarines and a strike bomb against surface targets. Dr. Robert Barker, Assistant to the Secretary of Defense for Atomic Energy, has testified that "the Navy and DOE are in the final stages of a phase 2a study to examine warhead candidates, including a variant of an existing B61."[30] The ASW ND/SB was scheduled to enter phase 3 development in July 1987,[31] but as of April 1988 DOD had not authorized proceeding to Phase 3. That step normally would entail the selection of a warhead candidate and a request for line item funding for construction of facilities. DOE has not requested funding for a new warhead for the ASW ND/SB, but it has requested increased funding for production of the B61.[32] If a new warhead were to be selected for the ASW ND/SB it is unlikely that it could be developed with sufficient confidence to be deployed under a CTBT. However, if the variant of the B61 is chosen as the warhead for this new bomb, then it is possible that development could continue and deployment of the ASW ND/SB could occur under a CTBT.

*The Antisubmarine Warfare Standoff Weapon (ASW SOW)*, also known as Sea Lance, would be fired from a torpedo tube of an attack submarine, rise to the surface, fly through the air like a missile, and return to the sea to attack an enemy submarine as a nuclear depth charge or homing nonnuclear torpedo. It is intended to replace the existing SUBROC missile, which functions in the same manner. Phase 2a studies on the ASW SOW have been completed, but in August 1986 the Navy decided to defer a decision on phase 3 development of the nuclear Sea Lance until further progress is made on the nonnuclear torpedo variant of the weapon.[33] That decision is now scheduled for June 1990;[34] no funds were requested for the ASW SOW warhead in the FY 1988 or FY 1989 DOE budgets. Funding for the ASW SOW warhead was requested in the FY 1987 DOE budget, before the Navy decision to defer the decision on phase 3 development, but Congress turned down the funding request. The House Armed Services Committee said the denial of funds for the ASW SOW weapon was merely a delay pending the Navy's decision on the weapons system,[35] but the Senate Armed Services Committee recommended cancellation of the nuclear Sea Lance.[36] There is no unclassified information available on the warhead candidates for the ASW SOW; therefore little can be said about the specific impact of a CTBT on this weapon system. A new warhead would seem to be precluded, however, if a CTBT goes into effect within the next few years.

*Strategic Earth Penetration Warheads (EPWs)* are designed to penetrate some tens of feet underground before detonating. The principle underlying EPWs is that a warhead exploded underground couples more of its energy to the earth than does one exploded in the air or on the earth's surface, creating more ground shock per kiloton of explosive yield.[37]

No currently deployed warhead could survive the stress of earth penetration. EPWs must be specially designed to endure tremendous forces of deceleration. EPW components must be much stronger than those of standard warheads, and are anchored inside a strong structural case for penetration. While this strengthening adds considerably to the warhead's weight, calculations indicate that this penalty may be more than offset by the effectiveness of the EPW compared to standard warheads against certain classes of targets.[38]

EPWs are in the feasibility study stage (phase 2) of development. At least one successful prototype test for proof of concept of an EPW has been done, and more testing is planned.[39] A successfully-tested prototype, however, would likely need more development before it would be suitable as a weapon. Assuming that an EPW is found to be a feasible concept, engineering development could start in perhaps two years.

If a CTBT were to take effect soon, several existing warheads could be used against buried, hardened targets instead of a totally new EPW, though with a degradation in effectiveness. (1) The W86 earth penetrator warhead was intended for use on the Pershing II when that missile was to be a short-range weapon. When it was decided to make the Pershing II into a longer-range missile, the W86 was no longer needed, as the missile would be used against different targets. Nonetheless, development of the W86 was completed.[40] It could be deployed for use in a tactical mode, but may not be suitable in a strategic mode because the yield is small. (2) Surface bursts from very high yield warheads would be effective against buried targets, but these warheads are extremely heavy, so a large missile might carry only one such warhead, compared to several EPWs. (3) An existing warhead might be modified for use in a limited earth penetration mode. With some means of controlling its speed of descent, such a warhead might withstand penetration into particularly soft media. Suitable modifications might not require nuclear testing but the target set for such a weapon would be limited compared to a more robust EPW that might be designed from scratch.

*Maneuvering Reentry Vehicles (MaRVs)* maneuver in the lower atmosphere to evade ABMs (which at present are sharply constrained by the ABM Treaty), enhance accuracy, or both. MaRVs are also in the feasibility study stage (phase 2) of development. A CTBT would not prevent either side from developing and deploying highly accurate MaRVs using existing warheads, but it would make the development of effective evasive MaRVs, which probably need more shielding than is on existing warheads, unlikely. These ideas are explained more fully below.

An accurate MaRV, unlike a ballistic RV,[41] would have inertial forces imposed on it as a result of modest turns, but existing warheads probably are sturdy enough to be used in an accurate MaRV without modifications that require nuclear testing.[42]

The purpose of an evasive MaRV would be to penetrate enemy defenses. In addition to being given maneuvering capability, an evasive MaRV would be designed to withstand high levels of radiation from potential attack. Therefore, a MaRV warhead could require more effective hardening than current warheads. Hardening could be improved without testing by adding shielding to the outside of the RV, but there could be weight and performance penalties in doing so. Nuclear effects testing,. however, could greatly improve weapon designers' ability to optimize shielding and other hardening measures with the least amount of weight and performance penalty.

If ABM deployments remain severely limited by the ABM Treaty, then MaRVs would not be needed for evading ABMs, and there could be a role for them solely to improve accuracy. Such MaRVs could be deployed with existing though not optimum warheads under a CTBT. However, confidence in such warheads would be below the current standard if shut-down production facilities for an existing warhead were restarted without a production verification test. If ABMs were unrestrained, or if one side wanted to guard against the other side leaving the ABM Treaty, then military planners might want MaRVs to evade ABMs, and possibly to enhance accuracy as well. Existing warheads would probably need extra shielding for this mission. Modifying an existing warhead for use in MaRVs would require extra X-ray shielding to compensate for uncertainties resulting from not testing, and neutron shielding outside the physics package to avoid interfering with its functioning. These two steps would add further weight and bulk to an evasive MaRV, increasing the attractiveness of such MaRV alternatives as additional ballistic RVs or penetration aids.

### Effects on the Development and Deployment of Future Generation Weapons

In general, the more significant a departure a weapon concept is from previous weapons, the more testing will be required to transform the concept into a workable weapon. Because the science as well as the technology of future generation nuclear weapons is poorly understood, there are no substitutes for nuclear testing in exploring their feasibility. Without nuclear testing, research relevant to advanced weapon concepts could continue with laboratory and computer experiments but it is highly unlikely that these concepts could be turned into weapons.

### Are the Effects of a CTBT on New Weapons Development to the Benefit of the United States?

Supporters and opponents of further limits on nuclear testing agree in some measure on the effects of a CTBT on the development and deployment of new weapons. The two sides part company, however, over whether a cessation in the deployment of new warheads by the United States and the Soviet Union is in the best interests of the United States. Both sides in the debate assert there will be greater implications for future weapons development than the evidence appears to support, in part for tactical reasons because neither side could press for its position on a CTBT if it conceded that the treaty would have little effect in that area. The arguments over whether the effects of a CTBT on new weapons development are to the benefit of the United States can be grouped into three areas: improving U.S. security, maintaining a technological advantage, and improving the safety and security of nuclear weapons.

***Improving U.S. Security.*** Testing supporters and testing opponents each assert that their position would be best for the security of the United States; testing supporters stress the benefits of maintaining and improving <u>deterrence</u> by developing new weapons while testing opponents stress the benefits of maintaining and improving <u>stability</u> by constraining development of new and potentially destabilizing weapons.

Testing supporters argue that U.S. nuclear weapons have deterred nuclear war as well as helped to prevent conventional war. Deterrence, in their view, requires continual development of modern nuclear weapons that both can threaten Soviet strategic warfighting targets and survive a nuclear war environment. Nuclear testing, its supporters conclude, is essential to development of these weapons and, therefore, essential to deterrence.

Testing opponents contend that the continual development of ever more sophisticated nuclear weapons by both sides is destabilizing and therefore threatening to U.S. security. This view holds that in a crisis either side might feel compelled to launch its nuclear forces first to gain maximum military advantage as well as to avoid losing them in a first strike by the other side. In this view, if new weapons development can be stopped or severely restrained, it would inhibit qualitative improvements in opposing nuclear forces and help stabilize the strategic nuclear balance between the two sides.

Testing opponents and testing supporters have differing views on the kinds of nuclear weapons needed for deterrence. Testing supporters believe that deterrence requires a diverse arsenal of modern weapons for attacking a wide range of Soviet military targets. They are concerned that without nuclear testing the deterrent value of U.S. nuclear forces will erode over time. For example, they argue that the Soviets can improve the survivability of their nuclear weapons and command and control infrastructure without nuclear testing; they can further harden missile silos, disperse their ICBMs on mobile launchers, improve their antiballistic missile systems, and improve their ability to destroy U.S. strategic submarines. A CTBT would limit severely U.S. ability to counter these kinds of Soviet actions. Testing opponents respond that a CTBT would limit Soviet ability to counter similar U.S. actions, which would be to the benefit of U.S. security.

Testing opponents view deterrence in a different light. In general, they believe that the capability to inflict unacceptable damage on an adversary in retaliation to an attack is sufficient for deterrence. In their view, the existing U.S. weapons stockpile more than suffices for this purpose and could continue to do so for many years to come, even if the number of warheads on each side is reduced sharply, as the current START negotiations are attempting to do. They do not foresee any technical advances that would avert massive damage from a retaliatory strike. They also assert that a CTBT would not preclude the United States from maintaining the reliability of its weapons stockpile.

Testing opponents believe that a CTBT would make its most useful contribution to stability by stopping or severely impeding deployments of nuclear weapons that entail radically new technologies and applications. Of greatest concern are weapons associated with strategic defenses (SDI). According to testing opponents, these weapons are extremely destabilizing because they can enhance a first-strike capability: one side could use its defensive weapons to destroy enemy space-based defenses, then attack enemy nuclear forces with its offensive weapons and use its defensive weapons to defend against a greatly weakened retaliatory strike. Testing supporters point out that the long-term objective of developing defensive systems is to reduce reliance on offensive nuclear weapons for deterrence and enhance security. They contend that the United States should not be denied the opportunity to exploit its technology advantage to explore the military potential of defensive weapons with the possibility that defense against missile attack may eventually replace or augment deterrence.

It should be noted that a nuclear test ban would have little effect on the type of SDI system proposed for deployment before the year 2000, the Phase I Strategic Defense System, because this system is entirely nonnuclear.[43] A test ban might, however, require more conservative design of the system's space-based components (e.g., more shielding) to compensate for a lack of nuclear effects tests. Nor would a test ban affect President

Reagan's stated vision for SDI, which is to develop nonnuclear defenses. It would, however, eliminate the possibility that futuristic nuclear weapons would be used in some SDI-derived system so long as the test ban were observed.

### *Maintaining a Technological Advantage.*

*Maintaining a Technological Advantage.* Supporters of continued testing believe that in general a high-technology competition with the Soviet Union favors the United States.[44] The United States developed the first nuclear weapons, was the first to detonate a thermonuclear device and to deploy MIRVs, accurate SLBMs, and lightweight cruise missiles. Testing supporters maintain that all these innovations were at least initially advantageous to the United States. Currently, the United States has more accurate missiles and warheads with higher yield-to-weight ratios than does the Soviet Union, so they can hold hardened targets at risk with a smaller missile throw weight. A CTBT would, in this view, give away the technological advantage the United States maintains through continued development of highly engineered weapons for which testing is essential.

Testing opponents believe that any advantage the United States might gain from continued testing would be transient, noting that the Soviet Union quickly emulated U.S. innovations in nuclear weapons. They believe that a CTBT would freeze the U.S. technological lead in warhead development. They agree that testing is crucial to development of sophisticated weapons and believe that it is to the advantage of the United States to limit warheads to those models currently in the U.S. and Soviet stockpiles.

The question of whether a CTBT would preserve or dissolve a U.S. advantage relates back to whether one views new weapons, particularly future generation weapons, as enhancing or reducing U.S. security. Testing supporters see the development of future generation nuclear weapons as potentially enhancing security and therefore believe it would be wrong to constrain the development of new advanced nuclear weapons--an area where the United States has a relative advantage--while not affecting areas of Soviet advantage (such as the survivability of targets).[45] As noted above, testing opponents see these weapons as reducing U.S. security.

Testing supporters believe that in addition to giving away the U.S. advantage in new weapons development under a CTBT, the United States would lose its current technical capability faster than the Soviet Union as many U.S. (but not Soviet) warhead designers would leave the laboratories for other pursuits. They believe that continued testing is needed to maintain a pool of skilled weapons designers who could make judgments about the reliability and effectiveness of weapons and respond if the Soviets break out of the treaty. During the 1958-1961 moratorium, according to DOE, Lawrence Livermore National Laboratory lost 10 percent of its weapons designers and diagnostic physicists and Los Alamos lost 30 percent; many more scientists indicated their intentions to leave if the moratorium had continued; and hiring became more difficult.[46]

Testing opponents argue that the relatively small loss of technical capability in the weapons laboratories would not be significant for national security and contend that the fear of losing funding is in fact a major reason why the weapons laboratories oppose more restrictive test bans.[47] They claim that if it were in the national interest to maintain teams of skilled warhead designers at the labs under a CTBT, then funding could be provided to do so. Scientists and engineers could be engaged in numerous activities related to warhead design that would be allowed under a CTBT, including inertial and magnetic fusion and modeling of complex hydrodynamic processes such as turbulent fluid flow. They could also design thermonuclear weapons that would remain untested and nuclear devices with near zero yield for which testing might be allowed under a CTBT. Furthermore, testing opponents claim that the Soviet Union would face difficulties similar to those confronting the United States in keeping their weapons scientists productively employed.

*Improving the Safety and Security of Nuclear Weapons.* The two areas of debate noted above relate to the costs or benefits of a CTBT for the United States vis-a-vis the Soviet Union. The debate over safety and security relates solely to whether the United States needs to improve those features of its warheads to reduce the chances of accidental dispersal of plutonium and the threat to warheads posed by terrorists and other unauthorized users. Testing supporters claim that safety and security improvements have been a major reason for weapons development over the past two decades and that continued improvements are needed, for reasons detailed in chapter 3. Testing opponents, on the other hand, hold that increased safety is not a sufficient reason to continue nuclear testing. They claim that the warheads most vulnerable to plutonium dispersal already use IHE and that replacing other warheads with those using IHE is not essential. The Trident II, for example, does not use IHE because the risk of plutonium dispersal from a warhead inside a submarine is quite low and has been judged not worth the extra weight of IHE, which would reduce the range of the missile. On the other hand, systems like the SRAM and the Navy's depth bombs do not now have IHE, but upgraded versions ready for phase 3 development would incorporate IHE and be safer.

Regarding permissive action links (PALs), testing opponents argue that all warheads requiring advanced PALs already have them. Furthermore, while some advanced PALs are part of the physics package of various warheads and their introduction involved nuclear testing, the United States could design PALs that are external to the physics package. This approach was used for many years and would not require nuclear testing.


**Effects of a Low Yield Threshold Treaty (LYTT)**


An LYTT would not constrain warhead development as much as a CTBT. The effect of an LYTT would obviously depend on the threshold. The main effect of thresholds on new warhead development is to push this development toward warheads that can be adequately tested at yields below the threshold. This section addresses the effects of an LYTT on weapons development and the strategic balance.

## *Effects on the Development and Deployment of New Current Generation Warheads*

An LYTT set at 1 kt would have almost the same effects as a CTBT on the development of new warheads. The main difference is that it would be possible to develop new warheads with yields in the vicinity of 1 kt under an LYTT. So-called neutron bombs have yields in that range, and therefore might be further developed and deployed in the future under a 1-kt threshold.[48] Some also believe it possible to develop pure fission weapons with yields as high as 100 kt by testing only at 1 kt.[49] There is, however, no apparent military value in such weapons compared to existing thermonuclear weapons of similar yield: they use much more fissionable material, so they would weigh several times as much and would cost much more.

A 20-kt LYTT would permit development of new thermonuclear warheads with yields somewhat above 20 kt with good confidence. However, designers' confidence in their new warheads would diminish the more the yield exceeded 20 kt. The reason is twofold. (1) The higher the yield of the secondary, the higher must be the yield of the primary to drive it. A 1-kt primary, for example, could not drive a 100-kt secondary. (2) To obtain confidence in a new warhead, designers need to test the secondary at least at partial yield to check how well the primary's energy couples to the secondary, the symmetry of the secondary's implosion, and the onset of fusion. Since the primary must be detonated to ignite the secondary, designers must test a warhead at a yield greater than the yield of the primary to have confidence in it. But as a warhead's total yield increases, the yield of the primary also increases, leaving a diminishing yield available under a 20-kt LYTT for partial-yield testing of the secondary. The information obtainable under a 20-kt LYTT is unlikely to provide sufficient information to develop warheads with yields in the range of current strategic systems. (Unclassified literature and discussions do not suffice for determining how much greater than the yield of the primary the test must be.)

New warheads of strategic significance, such as earth penetrators or MaRVs, might be deployed with confidence under a 20-kt LYTT. To destroy hard targets like ICBM silos, improved accuracy of delivery could compensate for reduced warhead yield.[50] To illustrate, a warhead with a yield of 300 kt and an average accuracy of 300 feet would have a 92 percent chance of destroying a target hardened to withstand a blast force of 5,000 pounds per square inch (psi). In contrast, a 30-kt warhead with an average accuracy of 150 feet would have an 89 percent chance of destroying a 5,000-psi silo. This accuracy is conceivable at ICBM ranges using MaRVs.[51] Similarly, an earth penetrator, which would transfer much more of the warhead's explosive energy to the earth than would a surface or air burst, would increase the silo-killing ability of low-yield warheads.[52]

### *Effects on the Development of SDI and Future Generation Weapons*

Less can be said about the effects of an LYTT on future generations of weapons, in part because such weapons are in an early stage and relatively little is known about what can eventually be achieved, and in part because much of what is known about such weapons is classified. A limited amount of information is available on the yields required for research on X-ray lasers and other nuclear directed energy concepts.

A nuclear-pumped X-ray laser uses energy from a nuclear explosion and converts a fraction of the energy into a focused beam of X-rays. Although the specifics of X-ray laser design and the yields required for them are classified, there are indications that (1) research on X-ray lasers can be conducted with tests with yields considerably below the current 150-kt threshold, but that (2) a threshold below 150 kilotons would hinder the development of effective weapons.

In general, the less advanced the X-ray laser technology, the larger the explosion required to generate lethal effects at a given distance.[53] Research on X-ray lasers is at an early stage, focusing on solving physics problems and determining whether X-ray lasers are feasible.[54] The reported testing pattern is consistent with these inferences. Of the five nuclear tests that Norris et al.[55] have identified from unclassified sources as

X-ray laser tests, the DOE-announced yields of 4 were between 20 and 150 kilotons (the highest yield category that DOE reports).

It is likely, however, that research could continue to be conducted at lower yields. Evidence for this is that one of the X-ray laser tests that Norris et al.[56] identified had an announced yield of under 20 kt, and an early X-ray laser experiment was conducted using X-rays from an effects test that had a yield of under 20 kilotons.[57] It also makes sense that yields in the same range as those currently needed to generate X-rays for effects tests would be needed to generate X-rays for X-ray lasers.[58] It is not yet clear if this research will lead to practical weapons under the current, relatively unrestricted testing regime.[59] The problems are more difficult and less likely to be solvable under a 20-kt limit. It is doubtful that an X-ray laser could be developed under a 1-kt threshold because it becomes much more difficult to generate X-rays of the required wavelength with small explosions.

There is less information on the yields required for testing other future generation weapons. Time magazine reported that the February 3, 1987, "Hazebrook" nuclear test was a 40-ton (.04 kiloton) test of the hypervelocity pellet weapon (the "nuclear shotgun").[60] Norris et al. reported a different test with an announced yield of below 20 kt as a test of the hypervelocity pellet weapon. Testing of hypervelocity pellet weapons at low yields is consistent with physical calculations. One kiloton is more than 80 thousand times the energy needed to accelerate 1 kilogram of material to a velocity of 10 kilometers per second--a velocity at which a small mass could destroy a reentry vehicle. Thus it would appear that experiments designed to develop a hypervelocity pellet weapon can be conducted well under a 1-kt testing limit.

From one point of view, many additional types of future generation weapons could, in principle, be developed with continued testing under even a 1-kt threshold.[61] Physical laws do not preclude weapons that concentrate most of their energy in microwaves, gamma rays, visible or infrared light, weapon debris, or neutrons. Many of these, in theory, could be made into directed energy weapons. If it becomes possible to achieve high efficiencies in converting the energy from nuclear explosions into focused beams and in narrow focusing of the beams, a 1-kt source could provide enough energy to make effective weapons. The military or strategic significance of such weapons is not clear and they would be of little use if the military could not determine their effectiveness, which could be difficult without testing in the atmosphere or in space. Thus, while new types of nuclear beam weapons might be discovered with testing below 1 kt, it is not clear if they would be attractive to the armed services under current testing restrictions. On the other hand, if testing above 1 kt were banned, research efforts would focus more on nuclear weapon possibilities below 1 kt.

### *Are the Effects of an LYTT on New Weapons Development to the Benefit of the United States?*

Any LYTT with a threshold below the current 150-kt limit would further constrain nuclear weapons development. As discussed above in connection with a CTBT, testing opponents see such constraint as benefiting U.S. security by improving stability and furthering arms control, while testing supporters see it as threatening U.S. security by reducing our ability to maintain and improve deterrence. Those arguments will not be repeated here. Suffice it to say that the lower the threshold of an LYTT the greater will be the constraints on weapons development and the more force those arguments have.

There are some additional issues to consider regarding weapons development under an LYTT. As noted above, a 1-kt LYTT would be very similar to a CTBT except that it would allow testing and deployment of new warheads with yields in the vicinity of 1-kt. Current generation warheads in this yield range are intended for use as battlefield weapons, particularly in the European theater. The Warsaw Pact has a major advantage over NATO in the numbers of troops and tanks stationed in Europe. Therefore, improving battlefield weapons may be viewed as more important for U.S. interests (e.g. NATO defense) than for the Soviet Union.

The United States and Soviet Union both have extensive arsenals of battlefield nuclear weapons. It is not clear whether there are significant near-term opportunities for improving these weapons, with the possible exception of enhanced radiation (ER) weapons. The United States has developed the neutron bomb, a battlefield ER weapon, but has determined that it is not currently in its interests to deploy it. Under an LYTT, with options for development of new weapons limited, there might be increased interest in revisiting the potential for neutron bombs. These warheads produce large amounts of lethal neutrons while minimizing blast effects and radioactive fallout. One use suggested for ER warheads is to disable enemy personnel inside tanks without causing extensive damage to the surrounding area, to counter the Warsaw Pact's major advantage over NATO in tanks. On the other hand, ER warheads could be used by advancing Warsaw Pact troops to disable defending forces without blocking forward passage.

Development of future generation weapons, such as the nuclear shotgun for which development could apparently proceed quite adequately under a 1-kt LYTT, is likely to be of more significance to the U.S./Soviet balance of power than marginal improvements in current generation battlefield weapons. The United States is generally believed to have a lead in future generation weapons. Under a 1-kt LYTT, the United States might be able to further exploit any technological superiority and make more effective use of tests at or below 1-kt for designing such weapons.

Our technological lead might favor the United States relative to the Soviet Union under a 20-kt LYTT as well. As discussed above, a considerable amount of new warhead development and deployment could proceed under a 20-kt LYTT, albeit at yields well below those of current strategic warheads. According to the Department of Defense, the United States is ahead of the Soviet Union in missile guidance technology.[62] To the extent that a 20-kt LYTT discouraged the development of high yield weapons, there would be an even greater premium on missile accuracy than there is today. In addition, the United States leads the Soviet Union in computer technology, which aids in assessing warhead performance without full yield testing,[63] so it might be able to develop higher yield warheads than the Soviet Union under the same testing constraints. In general, it seems reasonable that the country with more advanced technology could make more effective weapons under a testing threshold.

There will always be some uncertainty in estimating the yields of the rival nation's nuclear explosions. Therefore, under any LYTT it would be difficult or impossible to prove that tests just above the threshold were violations of the treaty. This aspect of threshold treaties can be seen as favoring the Soviet Union. Given the openness of U.S. society and the closed nature of the Soviet Union--glasnost notwithstanding--word of a clandestine test is more likely to leak out in the United States, and the Soviet Union could more readily install seismometers covertly on the other side's territory. As a result, the Soviet Union might be less deterred from testing clandestinely above the threshold, and would benefit more from an inadequate verification regime.

Soviet testing above the threshold of an LYTT can be deterred by strengthening the verification regime. In addition, the value of cheating can be reduced by choosing a threshold such that testing immediately above it offers relatively little military advantage. One kiloton may meet that criterion: it is substantially below the level where the majority of current primaries are believed to be tested.[64] Therefore, cheating on the margin is unlikely to assist development of strategic thermonuclear warheads. The extent to which there may be other military benefits to testing in the 1-3 kiloton range, for example in maximizing the performance of tactical warheads (including ER warheads) or developing future generation weapons, is uncertain. However, no warhead types are known to be made possible in that range that cannot also be developed through testing at 1 kt.

The situation at 20 kt is less clear. Testing for the onset of fusion and partial functioning of the secondary is crucial in developing a new or substantially modified thermonuclear warhead. Such testing could be done for lower-yield warheads under a 20-kt LYTT. (See pages 100-101.) For higher-yield warheads, in contrast, there would be inadequate yield available within the threshold for such testing. However, testing at yields slightly above the 20-kt limit could provide valuable information which could assist in the development of higher yield warheads. In that case, a 20-kt LYTT without extremely good verification provisions could present an advantage to the Soviet Union in that area. On the other hand, if Soviet primaries are larger than those used by the United States, they would not be able to learn as much from partial yield testing of secondaries up to a 20-kt limit as would the United States[65] and their potential for marginally exceeding the treaty limit might not be of great concern.

### Effects of a Quota Treaty on Weapons Development

A Quota Treaty, which would allow a small number of tests per year, would slow the development of new warheads and would most severely restrict weapons that require the most testing. Interest in a quota stems from a desire to permit a limited amount of testing for stockpile reliability and for effects while shutting off new warhead development. A Quota Treaty could also contain thresholds limiting the yield of permitted tests. The extent to which a Quota Treaty would slow warhead development would depend on the number of tests allowed, the thresholds set, and the relative priority of tests for warhead development in comparison with tests to maintain the reliability of existing warheads, effects tests, and physics tests.

There is currently competition for nuclear tests; tests are expensive and weapons designers have many devices that they would like to test. Under a Quota Treaty, with fewer tests allowed, this competition would intensify. DOE asserts that it currently conducts no more tests than are necessary, implying that all current tests have high priority.[66] Tests to confirm the reliability of stockpiled warheads or to confirm fixes in stockpiled warheads would have high priority if a single test could confirm the performance of thousands of deployed warheads, as that test would immediately enhance deterrence. Effects tests might have high priority because a single test can examine many devices for their resistance to the effects of nuclear explosions. Physics tests to aid the improvement of computer models might also have high priority because under a Quota Treaty there would be pressure to use increased computer simulation to reduce the number of tests required for each new nuclear device. New nuclear warheads have required on the average about six tests,[67] mostly developmental tests and a final production verification test,[68] before being deployed. Under a Quota Treaty, warhead developers would likely find ways to reduce the total number of tests required for new warheads. For example, they might make warhead designs more similar to previously tested models, and could conduct even more extensive nonnuclear testing and computer simulation. They could also "piggyback" more experiments on each test, conducting effects tests, weapons development tests, and physics tests on one explosion; however, compromises necessary to accommodate multiple purposes would likely result in less information than could be gained through separate tests. Assuming that DOE's testing program is already run at maximum efficiency, reducing the number of tests per weapon would have some cost in the design of the weapon, in the confidence in the weapon, or in time and money.

A quota set at a few tests per year could permit the United States to slowly certify warheads currently under development. A higher quota would permit faster warhead development, and would likely be opposed by CTBT advocates as doing too little. A declining-quota test ban, in which the quota is reduced periodically, could permit completion of warheads under development while slowing or halting development of additional warheads.

Nuclear directed energy weapons are likely to require many more tests than current generation warheads before they can become useful weapons,[69] so a Quota Treaty could well delay the development of such weapons for decades. Research on current NDEW concepts could, in principle, continue at a much slower pace that would depend on the level of the quota and the priority assigned to NDEW tests relative to other nuclear experiments. Several NDEW experiments can sometimes be conducted from one nuclear explosion, so some research could continue with a limited number of tests.[70]

109

# Notes

1. Brown, Paul S. Nuclear Weapon R&D and the Role of Nuclear Testing. Energy and Technology Review, September 1986: 6-18.
2. U.S. Department of Energy. The Need for Supercomputers in Nuclear Weapons Design. January 1986. 34 p. This source provides a detailed unclassified discussion of the role of computers in nuclear weapons design.
3. Testing of primaries is especially important because if a primary fails to produce the required yield, the secondary will also fail, perhaps giving the warhead a near zero yield.
4. Policy for Nuclear Research in the Strategic Defense Initiative. Signed by Caspar W. Weinberger, Secretary of Defense, Feb. 21, 1985, and by John S. Herrington, Secretary of Energy, Feb. 27, 1985. 1 p.
5. Brown, Paul S. Nuclear Weapon R&D and the Role of Nuclear Testing. Energy and Technology Review, September 1986: 15.
6. Taylor, Theodore B. Third-Generation Nuclear Weapons. Scientific American, v. 256, April 1987: 30-39.
7. Taylor, Third-Generation Nuclear Weapons, 39.
8. Electromagnetic Weapon Studies. Energy and Technology Review, June-July, 1986: 18-19. Also: Production of High Power Microwaves, ibid.: 20-21. The extent to which the research is tied to nuclear microwave weapons is not stated.
9. Statement of Donald M. Kerr. In U.S. Congress. House. Committee on Foreign Affairs. Subcommittee on Arms Control, International Security and Science. Proposals to Ban Nuclear Testing. Hearings, 99th Cong., 1st Sess. February 26; April 30; May 8, 14, and 15, 1985. Washington, Govt. Print. Off., 1985. p. 85.
10. Mark, Carson, in U.S. Congress. Senate. Committee on Foreign Relations. Nuclear Testing Issues. Hearings, 99th Cong., 2nd Sess. May 8; June 19 and 26, 1986. Washington, G.P.O., 1986. p. 57. See also Foley, Admiral S.R., Jr. Assistant Secretary of Energy for Defense Programs. Response to questions for DOE budget hearing. Subcommittee on Procurement and Military Nuclear Systems. House Armed Services Committee. February 19, 1986. p. 126.
11. The W52 was modified to change the high explosive system. Rosengren, J. W. Some Little Publicized Difficulties With a Nuclear Freeze. Washington, U.S. Department of Energy, October 1983. pp. 23-24.
12. Kidder, Ray E. Stockpile Reliability and Nuclear Test Bans: Response to J.W. Rosengren's Defense of His 1983 Report. Lawrence Livermore Laboratory Informal Report. February 1987. 19 p.
13. Foley, Response to questions for DOE budget hearing, p. 127.
14. Brown, Paul S. Nuclear Weapon R&D and the Role of Nuclear Testing. Energy and Technology Review. September 1986: 14.
15. Kidder, Ray E. Militarily Significant Nuclear Explosive Yields. Paper presented at the DOE sponsored Cavity Decoupling Workshop, Pajaro Dunes, California, July 29-31, 1985. p. 4.

16. Hydronuclear experiments involve a combination of high explosive, usually in a nuclear weapon configuration, and uranium or plutonium whose quantity is reduced far below the amount required for a nuclear explosion as the term is usually understood. The energy released by fission, while small, is not necessarily zero. During the 1958-1961 nuclear testing moratorium the United States conducted about 60 such tests with fission energy releases equivalent to less than the energy in one-hundredth of a pound of TNT and one test with a fission energy release equivalent to 0.4 pounds of TNT. Thorn, Robert N. and Donald R. Westervelt. Hydronuclear Experiments. Los Alamos National Laboratory. February 1987. 7 p. LA-10902-MS.

17. Statement of James Culpepper, Acting Assistant Secretary for Military Application, Department of Energy. In U.S. Congress. House. Committee on Appropriations. Subcommittee on Energy and Water Development. Energy and Water Development Appropriations for 1988. Hearings, 100th Cong., 1st Sess. Washington, Govt. Print. Off., 1987. Part 6, p. 829. (Hearings hereafter cited as FY88 House Energy and Water Appropriations Hearings.)

18. The phases through which warhead development proceeds are explained in Chapter 3.

19. James Culpepper, FY88 House Energy and Water Appropriations Hearings, p. 643.

20. Unclassified list of scheduled nuclear effects tests provided by DNA. See Table 9, pages 6-11 and 6-12.

21. U.S. Congress. House. Committee on Armed Services. Subcommittee on Procurement and Military Nuclear Systems. National Defense Authorization Act for Fiscal Years 1988/89 - H.R. 1748. Hearings, 100th Cong., 1st sess., Feb. 24 and 25, 1987. H.A.S.C. 100-12. p. 51.

22. Statement of Robert Barker, Assistant to Secretary of Defense for Atomic Energy, FY88 House Energy and Water Appropriations Hearings. p. 722.

23. Nuclear Weapons Databook, v. 2, p. 11.

24. Robert Barker, FY88 House Energy and Water Appropriation Hearings, p. 801.

25. James Culpepper, FY88 House Energy and Water Appropriation Hearings, p. 829.

26. Robert Barker, FY88 House Energy and Water Appropriation Hearings, p. 801.

27. Information for this point and the balance of this paragraph provided by John Harvey, Program Manager, Advanced Strategic Missile Systems, Lawrence Livermore National Laboratory, telephone conversations with Jonathan Medalia, July 10 and 23, 1987.

28. James Culpepper, FY88 House Energy and Water Appropriation Hearings, p. 829.

29. Robert Barker, FY88 House Energy and Water Appropriation Hearings, p. 801.

30. FY 1988 House Energy and Water Appropriations Hearings, p. 800.

31. FY88 House Energy and Water Appropriation Hearings, p. 643.

32. FY 1987 and FY 1988 DOE Congressional Budget Requests.

33. FY 1987 House Energy and Water Appropriations Hearings, p. 723.

34. FY 1987 House Energy and Water Appropriations Hearings, p. 643.

35. H.Rept. 99-718.

36. S.Rept. 99-331.

37. An EPW detonated 20 meters underground is reported to be 20 to 50 times more effective against buried targets than a warhead of the same yield detonated on the surface. John Morrocco, Defense Dept. Plans to Study Earth-Penetrating Nuclear Weapons. Aviation Week and Space Technology, June 8, 1987: 29.

38. Earth-Penetrating Weapons, Energy and Technology Review, June-July 1986: 4-5; and Weapons Chief: "We Can Leave Them in the Dust," [interview of Thomas Cook, Executive Vice President, Sandia National Laboratory, by David Lynch]. Defense Week, October 27, 1986: 9.

39. Earth-Penetrating Weapons. Energy and Technology Review, June-July 1986: 5.

40. Cochran, Thomas, et al., Nuclear Weapons Databook, v. 1, p. 292, 311.

41. A ballistic RV follows an unguided path through space and the atmosphere after being released from the missile. MaRVs require a mechanism to maneuver through the atmosphere, a guidance system to control the maneuvers, and extra sturdiness because of the stresses of maneuvering, making them larger and heavier than ballistic RVs of comparable yield.

42. Information provided by Gordon Guenterberg, Lawrence Livermore National Laboratory, to Jonathan Medalia, May 7, 1987.

43. U.S. Department of Defense. Strategic Defense Initiative Organization. Report to Congress on the Strategic Defense System Architecture, January 1988, p. 7-8.

44. See, for example, Lavoie, Louis. The Limits of Soviet Technology. Technology Review, November/December 1985: 68-75.

45. Statement of Admiral William J. Crowe, Jr., in U.S. Congress. Senate. Committee on Foreign Relations. Nuclear Testing Issues. Hearings. 99th Cong., 2nd Sess., May 8, June 19, and 26, 1986. Washington, Govt. Print. Off., 1986. p. 146-160.

46. U.S. Department of Energy. The Importance of Nuclear Testing. Briefing materials. 1987.

47. Statement of Dr. Josephine Anne Stein, before the California Legislature Senate Committee on Health and Human Services. Hearings on the Involvement of the University of California in the Testing of Nuclear Weapons. February 11, 1987.

48. Nuclear Weapons Databook, vol. 1, p. 28-29.

49. Taylor, Theodore B. Nuclear Testing is a Pandora's Box. F.A.S. Public Interest Report. Vol. 39, No. 10, December 1986: 10.

50. Specifically, a two-fold improvement in accuracy compensates for an eight-fold reduction in yield.

51. Klass, Philip. Pentagon Seeks Penetration Aids Action. Aviation Week and Space Technology, September 3, 1984: 45. Accuracy is measured as the radius of a circle centered on the target within which half of the warheads from a type of missile (or bomber) will fall. As another example, by

one report the Pershing II missile deployed in Europe uses a maneuverable reentry vehicle that has an accuracy of 65 to 130 feet. U.S. Congress. House. Committee on Foreign Affairs. Subcommittee on Europe and the Middle East. Committee Print: The Modernization of NATO's Long-Range Theater Nuclear Forces, by Simon Lunn. 96th Cong., 2d Sess. Washington, Govt. Print. Off., 1981. p. 21. The Pershing II is an intermediate-range ballistic missile; its RV is not packaged to withstand the heat generated by reentry at ICBM speeds. Nonetheless, the accuracy of its guidance system is independent of range, as it corrects its position by scanning the terrain below.

52. For example, depending on the type of soil or rock, a 1-kt weapon buried approximately 120 feet will have a crater radius roughly 2.5 to 3 times greater than if it is detonated on the earth's surface. Crater depth will also be much larger for the buried weapon. U.S. Departments of Defense and Energy. The Effects of Nuclear Weapons, third edition, compiled and edited by Samuel Glasstone and Philip Dolan. Washington, Govt. Print. Off., 1977. p. 255-256.

53. The effectiveness of an X-ray laser for a given yield of nuclear device depends on the efficiency with which the energy from the explosion is converted into focused X-rays divided by the square of the beam divergence angle. Thus either improving the efficiency or reducing the divergence angle (i.e., making the beam more narrow) would greatly reduce the yield of the explosion needed to make an effective weapon.

54. With regard to the X-ray laser, the Report to The American Physical Society of the Study Group on Science and Technology of Directed Energy Weapons concluded "This is a research program where a lot of physics and engineering issues are still being examined. What has not been proven is whether it will be possible to make a militarily useful X-ray laser." APS Study: Science and Technology of Directed Energy Weapons, April 1987: 5-6.

55. Norris, R.S., T.B. Cochran, and W.M. Arkin. Known U.S. Nuclear Tests July 1945 to 31 December 1987. Nuclear Weapons Databook Working Papers NWD 86-2 (Rev. 2A). Washington, D.C., Natural Resources Defense Council, January 1988. 60 p.

56. Norris, R.S., T.B. Cochran, and W.M. Arkin. Known U.S. Nuclear Tests July 1945 to 16 October 1986. Nuclear Weapons Databook Working Papers NWD 86-2 (Rev. 1). Washington, D.C., Natural Resources Defense Council, October 1986. 60 p.

57. Broad, William J. Star Warriors. New York, Simon and Schuster, 1985. p. 109.

58. Material has to be heated to a certain high temperature to generate X-rays of the necessary wavelength, and achieving these temperatures is easier with a larger explosion.

59. Even if successful, development is likely to take at least a decade. APS Study: Science and Technology of Directed Energy Weapons. op. cit.

60. Lemonick, Michael D. A Third Generation of Nukes. Time, May 25, 1987: 36.

61. Taylor, Theodore B. Third Generation Nuclear Weapons. Scientific American, v. 256, April 1987: 30-39.

62. Weinberger, Caspar W., Secretary of Defense. Annual Report to the Congress, Fiscal Year 1988. Washington, Govt. Print. Off., January 12, 1987. p. 247.

63. Weinberger, Annual Report to the Congress, p. 247.

64. Feiveson, Harold A., et al. A Low-Threshold Test Ban is Feasible. Science, vol. 238, Oct. 23, 1987: 457.

65. We do not know whether Soviet primaries are larger than U.S. primaries, but it may be reasonable to assume so given that the full yield of Soviet warheads tend to be larger than U.S. warheads and that in general larger secondaries require larger primaries. Furthermore, the capabilities that enable the United States to have higher yield to weight ratios in its warheads would be applicable to miniaturizing primaries.

66. U.S. Department of Energy. Policy Paper 5: Nuclear Weapons Testing. January 1987. p. 37.

67. Broad, William. U.S. Researchers Foresee Big Rise in Nuclear Tests, New York Times, April 21, 1986: 1.

68. Barker, Robert. Assistant to the Secretary of Defense for Atomic Energy, stated, testimony to the Senate Foreign Relations Committee Reprinted in Congressional Record, daily edition, August 7, 1986: S 10724.

69. Selden, Robert. Then-head of theoretical and computational physics at Los Alamos National Laboratory, telephone interview with Jonathan Medalia on January 9, 1987. Selden indicated that several tens of dedicated nuclear tests might be needed to develop an SDI-type weapon, though the precise number would depend on what is found in the course of these experiments.

70. For example, one early X-ray laser experiment was conducted on an effects test, and the next two experiments were conducted on the same dedicated test. Broad, William J. Star Warriors. New York, Simon and Schuster, 1985. p. 111, 116.

# 5

---

# Effects of More Restrictive Test Bans on Maintaining Confidence in the Warhead Stockpile

*Jonathan Medalia*

Confidence in stockpiled nuclear warheads can erode in two ways. First, nuclear warhead materials can decompose, corrode, and deteriorate in other ways over time. These effects of aging can lower the yield of a warhead or cause it to fail altogether. Effects of aging are monitored through inspection and needed remedies are normally made without nuclear testing.

Second, a warhead's design may be flawed. While warheads are extensively tested before certification, work done to develop new warheads or to explore warhead physics can reveal the existence of a flaw in previously-deployed warheads where none was suspected. Design flaws are normally found in the development process prior to certification.

Those who favor continued testing argue that nuclear testing is in certain instances the only way to evaluate flaws and the only way to have high confidence that a warhead will work properly after it has been repaired. As a result, they believe a CTBT "can be expected to severely undercut the capability of the U.S. to maintain a reliable stockpile."[1] In the view of testing supporters, if U.S. leaders were not confident that many U.S. nuclear warheads would work as intended, then our deterrent would be a bluff and our actions would be limited;

if the Soviets believed that many of our warheads would not work as intended, then the U.S. nuclear stockpile would not deter whether the warheads worked as intended or not.

Opponents of continued testing hold diverse views on the relation of confidence and testing. One view is that confidence could be maintained for decades at high levels without nuclear testing. Three other views agree that confidence would erode over time with a more restrictive test ban, but differ in how to cope with this situation. One view holds that a quota treaty, allowing one or two tests a year, would suffice for maintaining confidence. A second holds that, prior to entering a more restrictive test ban, the United States should replace its existing stockpile with warheads designed to be less sensitive to deterioration, and that if deterioration does occur, warheads could be remanufactured to original specifications or replaced with existing warheads capable of performing the required mission. A third is that some erosion of confidence is desirable: since a would-be attacking nation would have less confidence in the success of its attack yet would have to assume that the enemy's retaliatory strike could inflict substantial damage, the risk of nuclear war would decline.

A note on terminology is in order. The terms "reliability" and "confidence" are often used interchangeably in this debate. There is a distinction. Reliability is a technical term referring to the probability that a device will work as intended when called upon to do so. Confidence is a more subjective term, reflecting the trust one is willing to place in a device.

Ideally, the two should be related: when reliability is high, confidence should be high; when reliability is low, confidence should be low. Nuclear testing helps keep this relationship tight. A successful test gives evidence (though not conclusive evidence) that reliability is high, which in turn provides a basis for high confidence. A failed test strongly indicates that the warhead has unacceptably low reliability, which forces military and political leaders to place less confidence in the device until it is fixed. Under a more restrictive test ban, leaders might place high confidence in a warhead that in actuality was unreliable, or low confidence in a reliable one.

This chapter focuses on "confidence" rather than "reliability" for two reasons. First, each warhead type undergoes far too few nuclear tests, perhaps only one or two in its final design, to form a statistically useful measure of its reliability. Nonnuclear tests of warhead components, while conducted extensively, are inadequate by themselves for estimating total system reliability. Second, the contribution of warheads to deterrence ultimately depends on what the political and military leaders on both sides believe about how well warheads on both sides would work.

It is also important to distinguish between confidence in a warhead and confidence in a weapon system. A judgment of confidence in a weapon system takes into account the fraction of the force of that weapon system on alert at any time; the weapon's ability to survive attack before being launched; its ability to be launched (e.g., for the launch order to be transmitted to it), penetrate defenses, and reach the target; its reliability (to which the warhead contributes but one component), accuracy, and warhead yield; target characteristics (e.g., hardness); and so on. Political leaders must be concerned about confidence in a weapon system because that reflects a system's deterrent value. At present, with nuclear testing, warheads have a higher probability of functioning properly than do delivery systems, so warhead confidence is not a major factor limiting system confidence. Attention focuses on warhead confidence in the test ban debate because that is the aspect of weapon system effectiveness that a test ban might affect.

## Technology Background

U.S. nuclear warhead design has become exceedingly specialized over the years. Warhead design requirements include high yield-to-weight ratios (see page 60); variable yields;[2] special geometries to fit a wide array of delivery vehicles; enhanced resistance to radiation from detonation of U.S. and Soviet warheads and to the heat and vibrations of reentry; improved safety features to prevent unauthorized use, plutonium dispersal, or accidental detonation; tailored outputs to enhance radiation for neutron bombs or defensive warheads; and many more. A result of having a highly optimized design is that manufacturing tolerances are fine and minor flaws can cause warheads to fail.

At the same time, warhead materials deteriorate with age. Metals can corrode, some explosives and plastics can decompose, materials can interact, and radioactivity can accelerate these processes. Such effects can cause mechanical components to degrade over time.

Since the United States cannot detonate enough warheads for a useful statistical sample of each type deployed, and because of the importance and sophistication of these warheads, the Department of Energy (DOE) has a major program to examine and test warheads in the stockpile without nuclear testing. Randomly selected warheads are disassembled, materials are examined microscopically, and chemical analyses are performed. The performance of the chemical high explosive in a primary is checked without a nuclear explosion by replacing the fissionable material of the primary with inert material and detonating the explosive. The operation of all mechanical and electronic components is tested individually and in an integrated fashion.

If problems are found, fixes are almost always made without nuclear testing. Previous experience, derived in part from nuclear tests, can cast light on the nature of problems or the adequacy of fixes, and can permit minor changes to the nuclear components of warheads. Changes to many nonnuclear components such as fuzes can be checked without a nuclear test. If a problem affects most warheads of a particular type, the entire inventory of that warhead can be remanufactured to incorporate fixes, as has occasionally been done.

Nuclear testing complements this effort. Once a type of warhead has been deployed for a year or two, a warhead of that type is exploded.[3] This test provides information for stockpile confidence, assesses to some extent the effects of deployment and handling on the warhead, and seeks to determine if production has introduced flaws. Tests are used occasionally to verify changes made to correct problems identified by the nonnuclear testing program.

Since the mid-fifties we have had [up to 1971] five principal cases in which a nuclear test was an integral part of a corrective program for a nuclear weapon in our stockpile. These tests were conducted because it was necessary to change the design of the nuclear assembly system of a warhead in some important way ... because of a developing mechanical or metallurgical problem, or in order to meet stricter standards of nuclear safety.[4]

Sometimes, nuclear tests find problems that nonnuclear means and other nuclear tests fail to disclose.

In more recent experience, a primary failed when tested at -65°F, although nuclear tests conducted at normal temperature, calculations using our best computer codes, and nonnuclear hydrodynamic tests at the low temperature did not disclose major problems.[5]

Data on the use of nuclear testing in maintaining stockpile confidence has recently been declassified in two reports. One was prepared by the University of California's Scientific and Academic Advisory Committee (SAAC), which advises the president of the university on matters relating to the two weapon laboratories, which the university administers under contract to DOE.[6] The other report is by three officials at Lawrence Livermore National Laboratory (LLNL).[7] These reports provide the broadest unclassified coverage of problems that have affected stockpiled warheads. The data are summarized in a table in the LLNL report. That table lists the 15 deployed types of warheads, the problem or problems that affected it, and the purpose of conducting a nuclear test. Table 8 presents the LLNL table with modifications as noted on page 121.[8]

Supporters and opponents of continued testing interpret Table 8 in different ways. Supporters stress the importance of past tests for maintaining confidence. They conclude that of the 41 warheads that have entered the stockpile since 1958, 15 have required nuclear testing to find or fix problems. They point to the pace of testing: in the 27 years since the end of the moratorium, tests have been needed to find a problem in deployed warheads or to assure performance after a fix in 23 instances, nearly one instance (and more than one test) a year. Nor are problems confined to older systems: of the eight new

## Table 8

### Fifteen Warheads Deployed Since 1958 for Which Nuclear Testing Following Their Deployment Helped To Identify or Resolve Problems

| Warhead | Problem and Year of First Test to Resolve Problem | Years From Stockpile Entry to Test | Nuclear Testing Used to Identify or Evaluate Problem | Resolve Problem |
|---|---|---|---|---|
| Wxx | One-point safety concerns (1987) | >10 (1) | X | |
| W84 | Concern about marginal behavior at aged conditions (1984) | 1 | X | X |
| W79 | No practical manufacture of a complex part; different approach required altering the physics behavior (1982) | -1 (2) | | X |
| W80 | Failure at low temperature (1981) | 1 (3) | X | X |
| B61 | Concern about low-temperature performance (1981) | 2 (4,5) | X | |
| W68 | Degradation of high explosive (HE) (1980) | 10 | | X |
| W47 | Corroding fissile material (1963) | 3 | X | |
| | Vulnerability in simulated ABM environment (1962) | 2 | X | X |
| | Improvement of one-point safety (1963) | 3 | | X |
| W45 | Mechanical change of HE (1964) | 2 (6) | | X |
| | Performance under aged conditions (1964) | 2 | X | X |
| W52 | Replacement of HE because original wasn't safe enough for handling (1963) | 1 | X | X |
| B43 | Improvement of one-point safety (1962) | 1 | | X |
| | Performance under aged conditions (1962) | 1 | X | X |
| B28/W49 | Perf. under aged conditions (1962) | 4 | X | X |
| W44 | Improvement of one-point safety | | | X |
| | Performance under aged conditions | | X | X |
| W50 | Improvement of one-point safety | | | X |
| | Perf. under aged conditions (1962) | - 1 | X | X |
| B57 | Perf. under aged conditions (1962) | - 1 (7) | X | X |
| | Improvement of one-point safety | | | X |
| W59 | Improvement of one-point safety | | | X |
| | Performance under aged conditions (1962) | 1 | X | X |

## Notes to Table 8

This is the table presented in the LLNL report, with modifications as indicated in these notes. Date of stockpile entry for stockpiled warheads or major modifications of warheads was drawn from SAAC, p. 34; for warhead s no longer in the stockpile (W47, W52, and W59), this date was drawn from Cochran et al., Nuclear Weapons Databook, p. 8. Date of first test is from SAAC, p. 34. An entry is made under "years from stockpile entry to test" only when year of stockpile entry and year of test are known. A negative entry indicates that the first test for a specific problem was done before the warhead or modification entered the stockpile. Even though more than one test was conducted in some instances, date of first test is used instead of another date, such as completion of testing, because it is the only date available. Date of first test is of use in considering stockpile confidence issues because if a warhead does not need a confidence test within four years of its deployment, the odds are strong that confidence in it will remain high without testing it for many years.

(1) The Wxx's identity and year of stockpile entry are classified; Lawrence Livermore National Laboratory provided the figure ">10" for this report. A warhead is "one-point safe" if its high explosive (HE) can be detonated at any one point (in contrast to the intended detonation of the HE at multiple points simultaneously) yet produce no or negligible fission yield. One-point safe designs help protect warheads in peacetime against accidents and terrorists.

(2) The LLNL report indicates that the mod tested was the second one to be deployed (p. 18-19), which SAAC indicates was deployed in 1983 (p. 34).

(3) The LLNL report indicates that the mod tested was the first one (p. 21), which SAAC indicates was deployed in 1980. (p. 34)

(4) The mod tested was the mod 4. (Ray Kidder, Maintaining the U.S. Stockpile of Nuclear Weapons During a Low-Threshold or Comprehensive Test Ban. Lawrence Livermore National Laboratory, UCRL-53820, Oct. 1987, p. 18) SAAC indicates (p. 34) that the mod 4 was introduced in 1979.

(5) The LLNL report lists another problem for the B61 omitted from our table, replacement of high explosive with insensitive high explosive for safety. The SAAC and LLNL reports "differed about whether safety concerns actually required an HE change in the B61 system or merely made such a change desirable." (Paul White, Los Alamos National Laboratory. Apparent Discrepancies Between SAAC Report and LLNL Report: Nuclear Weapons Systems Requiring Post-Deployment Nuclear Testing, Jan. 22, 1988, p. 1. White's report was prepared in response to an earlier version of this chapter.) Our table uses the latter interpretation because the B61 had already undergone several modifications and DOE was at the time modifying selected warheads to use insensitive high explosive.

(6) While SAAC uses 1965 as the test date for the W45, Paul Brown, Assistant Associate Director for Arms Control, Lawrence Livermore National Laboratory, indicates the date was 1964. Interview with Jonathan Medalia, Jan. 7, 1988.

(7) While SAAC uses 1964 as date of stockpile entry for the B57, Paul Brown indicated that the actual date is 1963. Interview with Jonathan Medalia, Jan. 7, 1988. The Nuclear Weapons Databook (p. 8) also uses 1963.

or substantially modified warheads that have entered the stockpile in the 1980s, three have required testing for confidence, and of 23 problems for which testing was done, six occurred in the 1980s.

Opponents of continued testing believe that those favoring testing have inflated the need for confidence tests in three ways. First, several warheads have had a common problem, such as one-point safety or performance with aged tritium, so that one test could have identified or resolved a problem affecting several warhead types. Second, many problems for which testing was done occurred within a few years of the end of the 1958-1961 moratorium, when in this view some warheads entered the stockpile with inadequate testing. The resulting problems were found and fixed when testing resumed. Opponents of continued testing count these tests for development, not confidence. Disregarding these tests gives the result that of 53 warheads in the U.S. stockpile at some point since 1965 (including some introduced before 1965),[9] only six have had problems that nuclear testing was used to find or fix since 1965. Third, the table shows some tests conducted before the warhead or modification being tested entered the stockpile, indicating that these tests were developmental tests.

Opponents of continued testing see U.S. warheads as highly reliable once they have been in the stockpile for more than a few years. This table confirms only two instances in which a test was used to find, evaluate, or fix a problem in a warhead that had been deployed for more than four years.

## Points of Debate

### Is Nuclear Testing Necessary to Maintain Stockpile Confidence?

Test ban opponents hold that continued testing is essential to maintaining confidence for three reasons. First, and of greatest importance, they see an ongoing program of testing of all types as critical to developing and maintaining the expertise of weapons designers. Their skill is the key to maintaining confidence, for they continually evaluate the evidence as to the condition of stockpiled warheads and suggest fixes to problems. Their skill actually reduces the need for nuclear tests because they can often decide without nuclear testing when a problem in a stockpiled warhead needs fixing, and can often make the fix without nuclear testing.

Second, under a CTBT, many experienced weapon designers would leave the weapon laboratories and recruitment of new ones would be difficult because weapon development and maintenance would be scientifically uninteresting without the experimental check on ideas that only nuclear testing provides. As John Immele and Paul Brown of Lawrence Livermore National Laboratory state,

> The principal sources of motivation [for weapon designers] are an environment of scientific excellence and a sense of doing something important for the nation. A nuclear test ban would remove the principal scientific method from an environment of excellence and it would clearly signal that nuclear design was either unimportant or undesirable to national goals. A nuclear test ban would not only prevent us from training experienced personnel to maintain nuclear deterrence, it would also discourage high quality personnel from working in the area.[10]

The resultant loss of collective expertise at the weapons laboratories would contribute greatly to the erosion of confidence in the nuclear stockpile.

Third, the one or two[11] nuclear tests done each year for stockpile confidence are needed to certify reliability or to resolve problems. Admiral Sylvester Foley, Jr., Assistant Secretary for Defense Programs, Department of Energy, stated:

> Since 1958, 14 of the 41 weapon designs in stockpile or 34 percent of the weapons have required post-development nuclear tests to resolve problems. In three-fourths of these cases, the problems were discovered as the result of nuclear tests, and additional tests were required to confirm that the 'fix' was satisfactory.[12]

Since warheads are made to exacting tolerances, a design flaw or a defect that emerges with time in one or a few warheads of a particular type is often replicated in many or all warheads of that type. In this view, a single test could find or fix a defect that could affect confidence in and deterrent value of an entire class of weapons. As Robert Barker, Assistant to the Secretary of Defense for Atomic Energy, noted:

> ... without testing and with the inevitable age-related changes that occur in nuclear weapons, the situation may well arise in which one might believe that <u>no</u> weapons of a given type will work.[13]

Supporters of continued testing point to many instances in which problems were found only with nuclear testing, or in which fixes made without nuclear testing failed. One is the W52 warhead for the Sergeant tactical ballistic missile.[14] The warhead was extensively tested. In 1959, shortly before Sergeant's planned deployment, two accidental explosions involving chemical explosives of the type used in the W52 killed six people. In response, the Los Alamos Laboratory replaced the explosive in the W52 with another type. The 1958-1961 moratorium prevented a nuclear test to check the modified version, but Los Alamos had such high confidence in it that the new version was not tested for 17 months after the moratorium ended. When tested, the W52 produced only a small fraction of its expected yield. With the problem detected, designers fixed it and verified the fix with a nuclear test. Testing supporters see the W52 as a case in which a defect (here, a safety problem) emerged in a tested warhead, a fix made without nuclear testing created a dud despite designer confidence, a nuclear test was needed to detect the problem with the fix, and a nuclear test was needed to assure that the corrected design worked.

CTBT supporters see the foregoing arguments as overblown. They assert that nuclear testing might be nice to have in maintaining confidence in U.S. warheads but is never essential, for three reasons. First, they see warheads as highly reliable. Table 8 shows that warheads that have been in the inventory for a few years are very unlikely to develop problems for which nuclear tests would be performed under current practice. The average age at which nuclear tests have been done for warhead problems in Table 8 is over 2.3 years, and only two warheads have been tested for confidence after having been stockpiled for more than four years. In this view, if we would resist the temptation to rush every last warhead into stockpile just before a CTBT took effect, but used only already-deployed warheads, confidence would remain high.

Second, CTBT supporters hold that inspection and analysis have revealed all flaws or emerging defects in warheads.

> The assurance of continued operability of stockpiled nuclear weapons has in the past been achieved almost exclusively by non-nuclear testing . . . . It has also been rare to the point of non-existence for a problem revealed by the sampling and inspection program to require a nuclear test for its resolution.[15]

> In no case, however, was the discovery of a reliability problem dependent on a nuclear test, and in no case would it have been necessary to conduct a nuclear test to remedy the problem.[16]

Third, this position holds that alternatives to testing can fix any problems that do occur. Minor problems can be corrected confidently without nuclear testing. Problems with components outside a warhead's physics package do not require nuclear tests to fix. In the few instances where testing would be useful, confidence in a warhead type can be restored without testing by remanufacturing warheads exactly as they were before degradation occurred, using more robust warheads, or replacing failed warheads with warheads of existing types designed for a similar mission. These alternatives are discussed below.

CTBT supporters believe that scientists will join or stay at the weapon laboratories even if there is a complete ban on nuclear testing. Steven Fetter, a research fellow at the Center for Science and International Affairs, Harvard University, points out that many scientists should be motivated to stay at the laboratories under a CTBT.

> Much work would remain that is challenging and creative, laboratory equipment could still be first-rate, and the contribution to the national defense just as important. Scientists wanting a new challenge could move to non-weapon programs at weapon laboratories, where they would still be available for consultation about stockpile problems.[17]

Moreover, Fetter asserts, scientists could obtain nuclear warhead expertise without testing by studying warhead design theory, use of nonnuclear tests and computer simulations, studying data from earlier tests, and using inertial confinement fusion explosions for experiments on some aspects of fusion.[18] Finally, test-ban supporters would hold that less skill and fewer designers are needed to monitor a warhead's condition, make minor fixes, or to make major fixes by remanufacture, than to design new warheads.

### Can Remanufacturing Restore Confidence in a Warhead?

Sometimes, when a warhead design proves flawed or a defect emerges over time, all warheads of that type are replaced by newly manufactured warheads. The United States has done this at least five times.[19] When remanufacturing a warhead at present, the old design is modified to correct its flaws, to prevent known or suspected defects from reemerging over time, or both. In addition, the design may be altered to increase yield, improve safety features, and so on. The modified design may be tested if its designers feel the changes introduce an unacceptable level of uncertainty into warhead performance.

Some CTBT advocates recommend a variation to this approach to sustain confidence in the U.S. warhead stockpile under a CTBT regime. They argue that in the main inspection and nonnuclear testing will be adequate for identifying problems, and most problems will be correctable with adequate confidence. But in those instances--which would be rare in their view--where these measures failed to assure confidence that a particular warhead type would work as required, all the warheads of that type, or any defective components of that warhead type, could be remanufactured to the original design specifications. In this way, this position holds, nuclear testing and the uncertainties of not testing a modified design could be avoided while restoring confidence to what it was when the warhead was new. Richard Garwin notes that during a CTBT, we may find through inspection a problem with a warhead.

It rusts, corrodes, whatever, and we say, "How could we have been so stupid as to have made that mistake? Let's replace it with something better." The answer is "No!" You don't replace it with something better. You build it just the same way you built it before, but you go back and you replace that part and remanufacture it every two years at considerably higher maintenance costs than you would like.[20]

The expense of remanufacturing a warhead or warhead components to the original specifications would be at least in part offset by avoiding the cost of nuclear testing. In addition, the current stockpile confidence program occasionally incurs the cost of replacing old warheads with ones of modified design. If incurring the net expense, if any, of remanufacturing disposes of the stockpile confidence argument, CTBT supporters believe, it is money well spent.

Supporters of remanufacturing assert that effort is currently made to ensure that warheads can be remanufactured to identical standards. They point to a statement by Admiral Foley:

New warhead or bomb military characteristics[21] submitted by the Department of Defense for acceptance by the Department of Energy normally contain a requirement that the design, development, and production of the warhead (or bomb) be well documented and involve processes that to the extent possible allow replication of the warhead (or bomb) at a future date. Assuming, therefore, that vendor-supplied materials and components are still available at the time desired for remanufacture (and this will not necessarily be the case), the remanufacture of existing, well-tested warheads is possible.[22]

They also point to a requirement of the W87 warhead for MX:

It is desired that the warhead have an inherent endurance obtained as a result of design considerations that address: a maximum warhead lifetime, maximizing the ability to replicate the warhead at a future date, and maximizing the ability to incorporate the warhead in other weapon delivery systems. Therefore, the design, development, and production of the warhead must be well documented and involve processes that to the extent possible allow replication at a future date.[23]

To aid in remanufacture, components and materials could be stockpiled to maintain the same characteristics or chemical composition over time. Equipment used to produce materials that change over time could also be stockpiled.

CTBT opponents assert it would be impossible to have confidence that warheads of a current design remanufactured to their original specifications in the future would work without nuclear testing. As Roger Batzel, Director, Lawrence Livermore National Laboratory, and Robert Thorn, Deputy Director, Los Alamos National Laboratory, note:

> With respect to remanufacture, . . . experience has taught us that it is literally impossible to prevent changes in materials and workmanship quality and standards, or even specifications and working drawings, over an extended period of time. We have no way of knowing the impact of subtle changes in manufacturing practices, without performing a nuclear proof test to gain the confidence that the effects of seemingly harmless process changes were properly addressed.[24]

The LLNL report discusses difficulties of remanufacture. One example used is the remanufacture in the early 1980s of rocket motors for the Polaris A3 submarine-launched ballistic missile, the last of which had been built in 1968. This rebuild was done for Great Britain, which wanted the new motors to replicate the originals as closely as possible. As examples of the difficulty of remanufacture, the report cites the following:

> ...material [for insulating a rocket motor chamber] met the original specification requirements, but when it was used in a full-scale motor, a significantly different (more rapid) erosion rate occurred. A design change was required (increased insulator thickness) to achieve acceptable motor chamber insulation....
>
> One of the aluminum forgings used was Alcoa 7075-T6. It was the "same" alloy as used in the original manufacture according to Alcoa. In the interim, however, due to Alcoa facility changes and process improvements, the time from forging to quench had changed significantly. This resulted in forgings with higher internal stress characteristics, causing cracks in the forgings and subsequent rejections when the forgings were machined into adapters.[25]

CTBT opponents note that remanufacturing involves a severe penalty. By replicating the original flaws, this approach demands an endless remanufacture of that type of warhead. That is, if a particular flaw appears in a warhead type after ten years, all warheads of that type would have to be remanufactured every ten years or so. While that cost may be acceptable, the cost would be extreme if a defect emerged after only two years. And if after a warhead was introduced its design was discovered to be defective, then the warhead and any weapon system using it might have no military value.

## *Could More Robust Design Reduce the Incidence of Warhead Defects and Enhance Confidence in Remanufactured Warheads?*

Robustness refers to means of minimizing warhead sensitivity to design flaws or to defects emerging over time. For example, robustness might be increased by using more nuclear material or conventional explosive in primaries to reduce warhead sensitivity to deterioration that occurs with aging. Similarly, since tritium decays rapidly, robustness of warheads using tritium could be enhanced by replacing tritium more frequently.[26]

Some CTBT supporters acknowledge that without nuclear testing the armed services would not be totally confident that remanufactured warheads of current designs would work as intended. More robust warheads, they argue, can resolve this problem. These warheads would deteriorate more slowly than warheads of standard design, reducing the incidence of problems affecting confidence. Minor deterioration in more robust warheads, for example, would still leave confidence high that the conventional explosives would set off the primary and that the primary would trigger the secondary. These warheads would have to be designed and tested before a CTBT took effect. This position holds that more robust warheads could remain in the stockpile for a long time without problems, and when deterioration eventually set in they could be remanufactured without further testing yet with high confidence that they would work as intended. Ray Kidder, a physicist (although not a weapon designer) at Lawrence Livermore National Laboratory, stated:

I believe that robust nuclear weapons <u>can</u> be designed without excessive penalty in performance. That is, once designed, built, and proof-tested, their continued operability in stockpile can be assured with appropriate inspection, correction, and remanufacture programs without requiring further nuclear tests.[27]

These CTBT supporters charge that the potential unreliability under a CTBT of warheads of current design stems from a decision made sometime after 1971 that warheads should be designed on the assumption that nuclear testing would continue indefinitely.[28] With testing available to correct flaws that emerged, the laboratories could accept warhead sensitivity to minor deterioration in order to maximize yield-to-weight ratio, minimize use of special nuclear materials, and press the state of the art in other ways to obtain desired characteristics. Holders of this position would note a statement in the LLNL report:

... to increase range and decrease cost, we have often designed our warheads on the margin in terms of the yield-to-weight ratio or other especially important military requirements.... This approach is very cost-effective in terms of delivery systems, enabling aircraft and submarines to carry more missiles. On the other hand, it has meant that our warheads are more complex and thus more dependent on nuclear testing. We have paid for this system integration and efficiency with a warhead complexity that has increased our reliance on nuclear tests for certification.[29]

Put differently, this view holds that the United States has undercut its commitment to seek a CTBT, as stated in the Limited Test Ban Treaty and the Nuclear Nonproliferation Treaty, in order to build more efficient warheads.

Test ban opponents argue that the foregoing position really involves a massive program to develop and produce new robust warheads to replace all warheads currently in the U.S. stockpile. Not only would this program be costly and necessitate more nuclear testing (to certify the warheads before a CTBT took effect), but would also result in less efficient warheads. Currently, warheads are tailored to particular delivery systems. Under a CTBT, deploying simpler and more robust warheads could require the deployment of new delivery systems.

Alternatively, existing missiles might carry fewer warheads or lower-yield warheads to compensate for the reduced yield-to-weight ratio of less sophisticated weapons. Penalties resulting from more robust design that may not seem excessive to CTBT supporters would probably seem excessive to the armed services. As an example of the services' desire to avoid performance penalties, the Navy's W88 warhead for the Trident II does not use insensitive high explosive (IHE) in order to save weight and increase missile range.[30]

Fetter points out that the United States would not have to replace current warheads with more robust ones, but would only have to develop and test robust warheads.[31] Then, if the need arose during a CTBT to replace a deteriorated warhead, the design would be available and could be manufactured with high confidence--without further testing--that it would work. This approach would be much less costly than deploying such warheads. It would, however, leave policymakers open to fears that some warheads in the stockpile had underestimated or undetected flaws.

CTBT opponents note that more robust design cannot guard against all causes of failure. For example, one warhead failed at -65° F (see page 119) in part because IHE produces slightly different implosion characteristics at very low temperatures than at higher temperatures. Explosive testing of the IHE (without plutonium) at very low temperatures did not indicate that such temperatures would affect the nuclear explosion. Increasing the amount of fissile material or IHE in the warhead would not have made the warhead less susceptible to this problem.[32]

Another unforeseen failure occurred with the W68. The high explosive in that warhead deteriorated, giving off effluent gases that interacted with other warhead materials, particularly the adhesive used in the detonator. "Scientists feared that products from this reaction could interact with the detonator bridgewire and eventually cause the detonator to fail."[33] Adding explosive or adhesive would not have solved this problem.

The cost of replacing current warheads with more robust ones would hinge on the time available to design, test, and build the new warheads. If a CTBT were to take effect in a few years, replacing the entire current stockpile with more robust warheads would require a costly crash program. On the other hand, the Administration has stated that a CTBT is a long-term goal of U.S. policy. One means of preparing for a CTBT that would take effect many years from now would be for the United States to explicitly change the priorities of military characteristics of warheads (increasing the importance of robustness and replicability at the expense, perhaps, of yield or weight). In that case, more robust warheads could be phased in over many years, at much lower cost, as new warheads replaced old ones.

Some CTBT supporters believe that the current stockpile is sufficiently robust, and that CTBT prospects are hurt by those CTBT supporters who urge development of more robust warheads. Herbert York, former chief CTBT negotiator under President Carter, argues:

> My position ... has been and is that the stockpile is already robust enough to maintain deterrence during a test ban regime.... There is, in brief, no technical problem that needs to be fixed before we can have a test ban, and we should not act as if there were....
> The essential point is that in arguing the matter of stockpile reliability, those who believe it is already in hand should not at the same time push to have it fixed.[34]

These CTBT supporters would point to Table 8 in arguing that current U.S. warheads are highly robust. It shows that only two warheads had problems for which nuclear tests for stockpile confidence were performed once they had been in the stockpile for more than four years. They would hold that the DOD/DOE decision of 1982 to make explicit the requirement (the military characteristic; see note 21) for robustness[35] should make future warheads even more robust, as should the program that the laboratories started in 1981 of conducting about four weapon physics tests a year "intended specifically to develop an improved knowledge base that would lead toward a more reliable and reproducible stockpile under a test ban environment."[36]

## *Could Confidence Be Maintained by Substituting Effective Warheads for Deteriorated Ones?*

Advocates of a more restrictive test ban point out that many weapon systems or types of weapon systems carry more than one type of warhead. For example, there are five types of strategic nuclear bombs (B28, B43, B53, B61, and B83). Minuteman II uses the W56, and Minuteman III uses the W62 and W78. MX was originally planned to carry the W78, but the Air Force decided to deploy it with the W87. Trident II may carry the W76 and the W88.[37]   (Each individual missile is deployed with only one type of warhead.)   If confidence in a warhead declined, this position holds, that warhead could often be replaced with another warhead carried by the same or a similar weapon system.[38]

Supporters of continued testing point out that each warhead is tailored to a specific weapon system in terms of yield, weight, and so on, so that one warhead cannot substitute for another. According to Robert Barker, Assistant to the Secretary of Defense for Atomic Energy:

> In every case to date, the replacement [weapon] system has required a nuclear weapon different from that of the system it replaces. In some cases, physical dimensions alone preclude use of the older weapon. In other cases, existing warheads cannot survive the heat, acceleration, vibration and environmental extremes that a nuclear weapon will meet in the stockpile or during delivery. Even the yield requirement of the new system may be different from that of the system it replaces.[39]

The aggregate set of targets in the Soviet Union is composed of many types of targets:  hard and soft, point and area, strategic and conventional forces, military and urban/industrial. Warheads are allocated to targets depending on target characteristics, delivery system (including warhead) characteristics and availability, defenses, and target priorities. Supporters of continued testing assert that changing the mix of warheads on a missile type would compel changes in targeting for many other weapon types, using warheads less efficiently and destroying fewer targets.

Trident II illustrates the point. This missile is to carry 8 RVs.[40] Some Trident IIs will carry 8 W88s (reported yield: 475 kt), while others will carry 8 W76s (reported yield: 100 kt); accuracy for both is said to be 500 ft.[41] If the W88 developed problems under a CTBT and the United States could only substitute 8 W76s for 8 W88s, Trident II could destroy many fewer silos. Even if it carried 12 W76s as one payload option, as was reported to be the plan before the summit,[42] substituting 12 W76s for 8 W88s if the W88 developed problems would reduce U.S. countersilo capability.[43] Figure 4 shows the relation between number and yield of warheads, silo hardness, and silos destroyed.

If problems developed with the W76 under a CTBT, the W88 could be used. CTBT opponents would fault this substitution as well. W76 would be deployed to attack soft targets. Substituting 8 W88s for 12 W76s would reduce the number of soft targets Trident II could attack.[44] Substituting 8 W88s for 8 W76s would use more yield and special nuclear material than needed to destroy many targets, increasing cost and unintended damage for no military gain. If the W88 weighs more than the W76, as seems likely, replacing 8 W76s with 8 W88s would reduce missile range and submarine patrol area, increasing submarine vulnerability.

Supporters of continued testing would note that U.S. prompt counterforce capability in the 1990s will rest mainly on two warheads, the W87 (MX and Midgetman)[45] and the W88 (Trident II). Loss of confidence in one would jeopardize much of our hard target capability, and substituting other warheads would still leave this capability weakened. This position would also question, for reasons noted on page 128, our confidence in warheads newly manufactured to replace other warheads if we had to reopen the production line but could not test the new warheads.

Opponents of continued testing assert that the changes in capability that result from substituting one warhead for another are relatively modest, and are important only if one believes a nuclear war can be fought like a conventional war. If one believes that deterrence is insensitive to even substantial U.S.-

**Figure 4**

**Silos Destroyed by a Trident II Missile
With Selected Payloads**

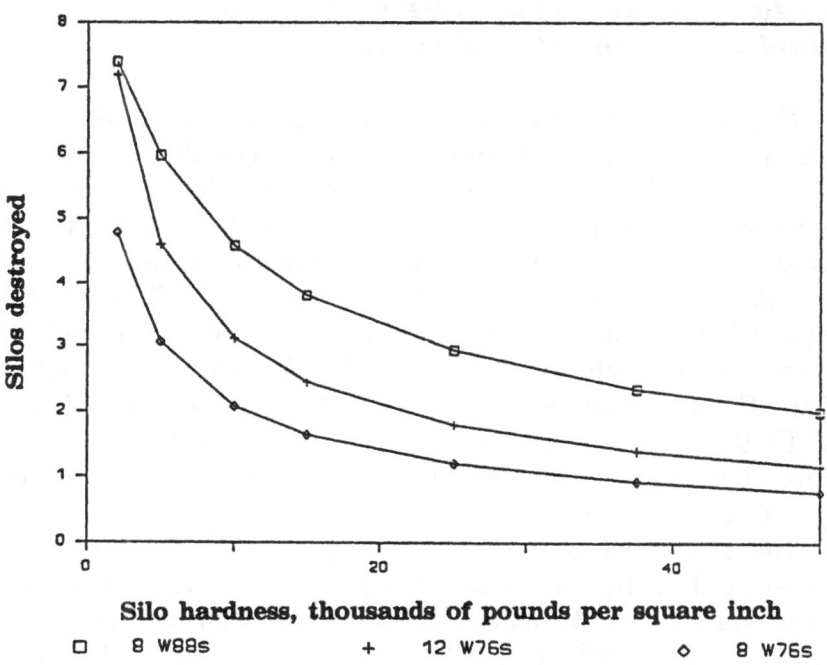

**Silo hardness, thousands of pounds per square inch**

□   8 W88s          +   12 W76s          ◇   8 W76s

Source:   Congressional Research Service.

Soviet disparities in nuclear capability, then substituting one warhead for another would not affect deterrence.

## Would the Effects of a CTBT on Confidence Disadvantage the United States?

Supporters of continued testing fear that a CTBT would reduce confidence in stockpiled nuclear warheads. Their worst fear is that all warheads of one type could deteriorate to the point where an entire weapon system using that warhead would become militarily useless, with disastrous results for U.S. security. For example, if Midgetman and MX were fully deployed in the late 1990s and a flaw emerged in the W87, the counterforce potential of the U.S. ICBM force could plummet. If the Soviets experienced similar problems, such as with their SS-18 ICBM warheads, they could more easily hide the problems than could the United States because of the greater openness of U.S. society.

Deterrence, in this view, rests on predictability. If each side perceives that the other side's weapons work, the temptation to raise demands or to attack in the hope that the other side cannot retaliate will remain low. This perception is the core of deterrence: the belief that weapons will work is more important to deterrence than whether they actually work. While a program of disinformation, with nuclear programs pursued only to the extent needed to support it, could in theory deter on the cheap, for a superpower that course is fraught with peril. The surest way to deter, and the only way to obtain usable military capability, is a vigorous program of developing and maintaining strong military forces. As part of this effort, in this view, nuclear testing for the maintenance of confidence must continue.

CTBT opponents are particularly frustrated by some CTBT advocates who oppose nuclear tests but support more testing of nonnuclear military systems like armored infantry vehicles and air defense guns. These CTBT advocates assert that our soldiers cannot depend on weapons that are not tested realistically. CTBT opponents argue that, to be consistent, these

CTBT advocates should support testing of nuclear warheads, which are more complex than most nonnuclear systems, to assure confidence in warheads.

CTBT supporters see the confidence issue as phony. They believe it was raised by CTBT opponents only as great progress in seismology rendered verification an unconvincing rationale for opposing a test ban. Even if confidence is addressed on its own terms, CTBT supporters believe that the argument does not provide a rationale for opposing a test ban. Many supporters see a CTBT as having little effect on the strategic balance. In this view, the United States and Soviet Union could maintain stockpile confidence at a high level for decades by nonnuclear testing, perhaps augmented by remanufacture, robust warheads, and substitution.

Some CTBT supporters go further and argue that a CTBT would promote stability by causing each side's confidence in its warheads to decline at about the same rate:

> There is no adequate substitute for such reliability testing if the military is to maintain confidence in its weapons. Thus a ban on all testing would mean that, over a long period of time, there would be a gradual deterioration of confidence in the reliability of nuclear weapons . . . A gradual reduction in stockpile confidence would discourage either side from contemplating a preemptive "first strike" against the other's nuclear weapons. First strike weapons must perform precisely and reliably . . . the nation launching a retaliatory strike would not need the same level of confidence.[46]

Both groups of CTBT supporters therefore hold that we should close the debate on confidence and look to a CTBT for the advantages it offers in slowing the arms race, slowing nuclear proliferation, strengthening the regime of arms control, and reducing the risk of war.

CTBT opponents fear that U.S. warheads would deteriorate faster than Soviet warheads under a CTBT for two reasons. First, U.S. warheads are more sophisticated than Soviet warheads and consequently allow less tolerance for change, so that a small deterioration is more likely to harm U.S. than Soviet warheads. As a DOE briefing on test ban issues noted,

> Nuclear testing appears to be more important to the U.S. than to the Soviet Union . . . . We rely more on high technology and on optimized warhead characteristics in our nuclear warhead design . . . In a no-test environment, Soviet missile throw-weight and volume advantages would permit the Soviets to fall back on previously-tested, heavier, and relatively simpler warhead designs, which generally should be more reliable and durable.[47]

Similarly, Brent Scowcroft, the chairman of the President's Commission on Strategic Forces, and two other members of the commission argued that "with our greater reliance on quality than on quantity, our confidence in our weapons might well deteriorate much faster than the Soviets' confidence in theirs."[48] Second, the Soviets could keep experienced people at weapons laboratories, whereas U.S. weapon designers would leave for other opportunities, giving the Soviets an edge in maintaining warheads.

CTBT supporters counter that we cannot assume that U.S. warheads would deteriorate faster than Soviet warheads under a CTBT. We know very little about Soviet warhead design.[49] Even if they are not as sophisticated as ours, that does not mean that Soviet warheads are more robust, or their design more conservative, than ours. The Soviet state of the art might be less advanced than that of the United States; yet they could be pushing the edge of their state of the art just as hard as we are, but with less sophisticated warheads.[50] Regarding the argument that the Soviets could retain people at weapon laboratories, the pro-CTBT position holds that some designers would stay at U.S. laboratories and that fewer designers would be needed to maintain confidence in stockpiled warheads than to develop new ones. Furthermore, in this view, a CTBT would over the longer term undercut the nuclear design competence of U.S. and Soviet designers alike.

CTBT opponents respond that if the United States knows little about Soviet weapon design, it should not rest its security on an unknown probability that the United States and Soviet Union would retain similar confidence in their warheads during a CTBT. That course could put deterrence at risk. Instead, this nation should continue nuclear testing in order to maintain high confidence in its warheads.

CTBT supporters would note that Table 8 shows only two warheads for which tests were performed for confidence on warheads that had been in the stockpile for more than four years. The U.S.S.R. could hardly do better. In this view, the historical experience provides no basis for alleging that U.S. warheads will deteriorate more or faster than Soviet warheads, as long as all warheads in the stockpile when a CTBT takes effect have been deployed for several years.

CTBT opponents respond that the W87 and W88 will comprise all U.S. ability in the 1990s to destroy rapidly the hardest targets. Yet they will not have been deployed long in the early 1990s: W87/MX, 1986; W88/Trident II, 1989; and W87/Midgetman, 1992. Since many defects are found within a few years of a warhead's stockpile entry, a CTBT entered in the early 1990s would impose a high risk of leaving entire weapon systems crucial for deterrence without warheads in which our leaders can place confidence. They would also point out that since six warheads have had confidence tests in the 1980s, as Table 8 shows, we cannot assume that no problems will emerge with the W87 and W88.

CTBT supporters typically oppose counterforce capability. Many would see the reduction in confidence in counterforce warheads as a desirable outcome of a CTBT negotiated by the next Administration. They would also point out that the United States has a qualitative lead in prompt counterforce capability, given that U.S. ICBMs and SLBMs are reported to be more accurate than their Soviet counterparts.[51] If a CTBT would slow or stop Soviet acquisition of counterforce capability, that would further reduce the risk of nuclear war.

The irony is that CTBT supporters view the substantial warhead deterioration that CTBT opponents believe might occur under a CTBT as quite acceptable, while CTBT opponents view the minor warhead deterioration that CTBT supporters believe might occur under a CTBT as jeopardizing deterrence.

## Effects of Alternative Test Bans
## on Stockpile Confidence

### Low Yield Threshold Treaty

Proponents of an LYTT seek to achieve most of the hoped-for gains of a CTBT while satisfying concerns that under a CTBT the United States could not adequately monitor low-yield Soviet testing or maintain confidence in its warheads. This section deals with the latter concern.

Would a threshold of around 20 kt suffice for maintaining confidence? Most confidence problems involve primaries, and, while DOE will neither confirm nor deny this figure, advocates of an LYTT assert that yields of primaries are below 20 kt.[52] According to former Secretary of Defense James Schlesinger, 15 kt is the minimum testing yield needed "to assure stockpile reliability;" former Secretary of Defense Harold Brown places the yield at 10 kt.[53] A 20-kt LYTT would also permit some partial-yield testing of secondaries.

In contrast, DOE argues that maintaining confidence in stockpiled warheads requires higher thresholds. According to Donald Kerr, former Director, Los Alamos National Laboratory, a threshold of "a few tens of kilotons" is required "for the purpose of assuring weapon performance. Even with a threshold at this level, some elements of the stockpile might eventually be in jeopardy."[54] Similarly, a DOE policy paper notes, "Only by testing at yields up to 150 kilotons are we able to validate performance of the fusion stage of many strategic weapons."[55] Weapon laboratories also fear that a 20-kt LYTT could reduce the expertise of designers of warhead secondaries, in turn reducing their ability to maintain confidence in secondaries in stockpiled thermonuclear warheads.

The two positions are not as far apart as they appear. Both sides agree that few nuclear tests are needed to resolve confidence problems, that a 20-kt LYTT would permit resolution of most of these problems, and that higher thresholds would allow resolution of the few remaining problems. The dispute comes down to the penalties involved in suboptimal resolution of the few problems that cannot be completely resolved with testing below 20 kt.

Attention has recently turned to the possibility of maintaining stockpile confidence under a 1-kt LYTT. This issue assumes particular importance in light of efforts in the House of Representatives to legislate such a ban. Ray Kidder concludes that confidence could be maintained under this ban by remanufacturing warheads,[56] using several steps he sees as feasible: making warhead specifications and their production procedures detailed enough for remanufacture; maintaining the availability of needed materials by stockpiling them or the equipment used to make them, or by assuring that the requisite production lines remain open; maintaining the skills of the people who produce warheads by remanufacturing warheads on a continuing basis; and retaining scientists and engineers at the weapon laboratories and maintaining their skills by continued nuclear testing and related projects.

Kidder stresses the advantages to maintaining confidence of a 1-kt LYTT as compared to a CTBT. He states, "There is no doubt that an interesting and vigorous program of theoretical, experimental, and computational nuclear weapons research could be conducted within the limits of [an LYTT] that would engage the interest and maintain the skills of nuclear weapon designers."[57] This program would include:

" ... all one-point safety tests."[58] These tests determine if a warhead will produce a nuclear yield if the high explosive is detonated at a single point, as might happen in an accident. Many confidence tests have been done for this purpose. Since a successful one-point safety test has no nuclear yield, such tests could be done under a 1-kt LYTT.

Testing of most primaries and single-stage warheads at full unboosted yield. In boosting, deuterium-tritium gas undergoes fusion by burning during the explosion of a primary. The resulting neutrons enhance fission, increasing the warhead's yield by a large amount.[59]

"A very broad spectrum of nuclear weapons research ..."[60] Examples were deleted from the report; unclassified examples are computer simulation, inertial confinement fusion, hydrodynamic tests (which use no fissile material), hydronuclear tests (which use only enough fissile material to produce the explosive yield of a few pounds of TNT), and 1-kt tests.

Kidder believes that this program would enhance stockpile confidence by helping retain technical personnel at the laboratories and maintain their skills, and by permitting tests to resolve some stockpile confidence problems. Several officials at Livermore take issue with Kidder. They maintain that we could not be confident without testing at yields much higher than 1 kt that remanufactured warheads would work, for reasons noted on pages 5-17 and 5-18.

They point out several constraints that a 1-kt LYTT would impose on confidence. First, boosting is crucial to modern warheads but cannot be tested at 1 kt, as Kidder recognizes.[61] Yet "boosting can only be certified reliably with full-scale testing ... boosting is a subtle process and small, seemingly insignificant, changes can cause it to fail; thus nuclear testing is essential."[62]

Second, while one-point safety tests can be conducted within a 1-kt threshold, "if a weapon fails a one-point safety test, the weapon design may have to be changed and this will require much higher yields."[63]

Third, expertise needed for confidence would diminish under a 1-kt LYTT. As with most engineering tasks, warhead designers specialize in various areas: high explosives, metallurgy, electronics, implosion physics, boosting, linkage between primary and secondary, fusion, etc. As the testing threshold is lowered, moving from the current regime to a 20-kt LYTT, a 1-kt LYTT, and a CTBT, fewer specialists can conduct the tests crucial to their part of warhead design. Miller and Brown state: "We believe that the nuclear test yields that are needed to certify nuclear components are the same yields needed to maintain the relevant scientific skills."[64]

Finally, CTBT opponents doubt that testing under a 1-kt threshold would provide sufficient motivation for scientists and engineers to remain at the weapon laboratories.

## Quota Treaty

The rationale for a Quota Treaty is to permit the maintenance of confidence (responding to a main objection of testing advocates to a CTBT) while achieving the main goal of CTBT supporters, constraining the development of new weapons. The case for a Quota Treaty is that the United States conducts one or two confidence tests a year on stockpiled warheads,[65] so an agreed limit near that level on the total number of nuclear tests per year would not impinge on the U.S. stockpile confidence program. Some CTBT supporters favor a Quota Treaty as a steppingstone to a CTBT. In this view, the only politically feasible approach to a CTBT is step by step, building upon the Limited Test Ban Treaty of 1963 and the Threshold Test Ban Treaty of 1974. A Quota Treaty would be another step in this direction.

A Quota Treaty would be opposed by two groups, those who believe it would allow too little nuclear testing and those who believe it would allow too much. CTBT opponents assert that all current tests bear on confidence by improving the ability of weapon designers to understand the behavior of nuclear weapons. In this view, one reason that few tests are currently used to discover warhead confidence problems or to check fixes is that warhead designers are able to judge without nuclear testing when a problem will affect confidence and when a fix can be confidently made without testing. Moreover, stockpile confidence tests are relatively unimportant in helping designers enhance their skills. These tests, while instrumented, make little or no attempt to understand the limits of warhead performance, so are the least instructive type of test for warhead designers. Thus, a sharp reduction in the number of tests would reduce the expertise of weapons designers that now allows the United States to maintain stockpile confidence with only one or two tests a year for that purpose. In that case, minor instances of deterioration that would now be judged acceptable might be judged excessive, so that the quota might permit too few tests to maintain confidence.

Some CTBT supporters might oppose a Quota Treaty. Those who believe that any decline in confidence resulting from a CTBT is good might oppose a Quota Treaty on grounds that it would permit maintenance of confidence. Others who believe that confidence can be maintained without nuclear testing could oppose a Quota Treaty as being an unnecessary concession to testing proponents.

## Notes

1. Rosengren, Jack. Stockpile Reliability and Nuclear Test Bans: A Reply to a Critic's Comments. Prepared for Office of International Security Affairs, U.S. Department of Energy. No. RDA-TR-138522-001. R&D Associates, Arlington, Va., November 1986. p. 1.

2. Cochran, Thomas, William Arkin, and Milton Hoenig. Nuclear Weapons Databook. Volume I: U.S. Nuclear Forces and Capabilities. Natural Resources Defense Council. Cambridge, Mass.: Ballinger, 1984. p. 38.

3. Foley, Admiral Sylvester, Jr., Assistant Secretary for Defense Programs. Department of Energy. Responses to Questions for DOE Budget Hearing, submitted by Rep. Stratton, February 19, 1986. In U.S. Congress. House. Committee on Armed Services. Subcommittee on Procurement and Military Nuclear Systems. Department of Energy National Security Programs Authorization Act for Fiscal Years 1987 and 1988. Hearings on H.R. 4526 [H.R. 4428], 99th Cong., 2d Sess. Washington, U.S. GPO, 1987. p. 127.

4. U.S. Congress. Senate. Committee on Foreign Relations. Subcommittee on Arms Control, International Law and Organization. Hearings: Prospects for a Comprehensive Nuclear Test Ban Treaty. Testimony of Carl Walske, Assistant to the Secretary of Defense for Atomic Energy. 92d Cong., 1st Sess. Washington, U.S. GPO, 1971. p. 106.

5. Foley, Sylvester. Responses to Questions for DOE Budget Hearing. U.S. Congress. House. Committee on Armed Services. Subcommittee on Procurement and Military Nuclear Systems. Hearings on H.R. 4526 [H.R. 4428]. p. 126.

6. Scientific and Academic Advisory Committee. Nuclear Weapons Tests: The Role of the University of California-Department of Energy Laboratories. A Report to the President and the Regents of the University of California. July 1987. 37 p.

7. George Miller, Associate Director for Defense Systems, Paul Brown, Assistant Associate Director for Arms Control, and Carol Alonso, Deputy Leader, "A" Division. Report to Congress on Stockpile Reliability, Weapon Remanufacture, and the Role of Nuclear Testing. Lawrence Livermore National Laboratory, UCRL-53822, Oct. 12, 1987. 59 p.

8. While SAAC presents a similar table, the LLNL table was used for two reasons. (1) The LLNL table has better data, in part because the weapon laboratories, which provided data for SAAC, found additional information after the SAAC report was completed that was used in the LLNL report. (2) A

table synthesizing the SAAC and LLNL tables is misleading. It emphasizes disparities between the two, of which there are many. Of 27 problems identified in one table or the other, only 19 were noted in both, and for these 19, the two tables agreed on the purpose of the test in 13 instances and disagreed in six. The disparities, however, are minor, arising largely from differing ways of interpreting the same data.

9. Cochran et al., Nuclear Weapons Databook. Table 1.4, U.S. Nuclear Warheads (1945-Present), p. 7-9. Most of the difference between "the 41 warheads that have entered the stockpile since 1958" (referred to in the preceding paragraph) and the "53 warheads in the U.S. stockpile at some point since 1965" arises because 11 warheads in the stockpile at some point in 1965 or later entered the stockpile in 1958 or earlier. Ibid.

10. Immele, John, and Paul Brown. A Commentary on "Stockpile Confidence During a Nuclear Test Ban." International Security, forthcoming 1988. Manuscript, p. 14.

11. Broad, William. U.S. Is Committed to Nuclear Tests. New York Times, Oct. 18, 1987: 1. A publication of Lawrence Livermore National Laboratory indicates that "Laboratory weapons scientists conduct about two stockpile confidence tests each year." Nuclear Weapon R&D and the Role of Nuclear Testing, Energy and Technology Review, September 1986: 11.

12. Response by Admiral Foley to Representative Markey, April 17, 1986. Cited in Ray Kidder, Maintaining the U.S. Stockpile of Nuclear Weapons During a Low-Threshold or Comprehensive Test Ban. Lawrence Livermore National Laboratory, UCRL-53820, Oct. 1987, p. 4. The test of the Wxx in 1987 has changed the data to 15 of 41 designs.

13. Barker, Robert. Letter to the editor, Physics Today, December 1983: 13. Emphasis in original.

14. This example is drawn from Jack Rosengren. Stockpile Reliability and Nuclear Test Bans: A Reply to a Critic's Comments. R&D Associates, Nov. 1986: 22-24; and John Immele, Deputy Associate Director (Nuclear Design), Lawrence Livermore National Laboratory, personal communication, June 1987.

15. Letter from Norris Bradbury, former Director, Los Alamos National Laboratory, J. Carson Mark, Head, Theoretical Physics Dept., Los Alamos National Laboratory, and Richard Garwin, consultant to Departments of Defense and Energy and former member, President's Science Advisory Committee, to President Carter, dated August 15, 1978; reprinted in U.S. Congress. House. Committee on Foreign Affairs. Subcommittee on Arms Control, International Security and Science. Hearings and Markup: Proposals to Ban Nuclear Testing, on House Joint Resolution 3. 99th Cong., 1st Sess. Washington, U.S. GPO, 1985. p. 214. Emphasis in original.

16. Bethe, Hans, and Norris Bradbury, Richard Garwin, Spurgeon Keeny, Wolfgang Panofsky, George Rathjens, Herbert Scoville, and Paul Warnke, letter to Rep. Dante Fascell dated May 14, 1985; reprinted in DeWitt and Marsh, Weapons Design Policy Impedes Test Ban: 11.

17. Fetter, Steven. Stockpile Confidence Under a Nuclear Test Ban. International Security, Winter 1987/1988: 146.

18. Ibid., p. 146-147. For more on means of obtaining nuclear design information without nuclear testing, see pp. 82-84.

19. Rosengren, Jack. Some Little-Publicized Difficulties with a Nuclear Freeze. Marina del Rey, CA: R&D Associates, 1983. Sponsored by the Office of International Security Affairs, U.S. Department of Energy. p. 16-17. Reprinted in U.S. Congress. Senate. Committee on Foreign Relations. Hearings: Nuclear Testing Issues. 99th Cong., 2d Sess. Washington, U.S. GPO, 1986. p. 178-180; and Rosengren, Jack. Stockpile Reliability and Nuclear Test Bans: A Reply to a Critic's Comments. Prepared for Office of International Security Affairs, Department of Energy. R&D Associates, RDA-TR-138522-001, November 1986: 16-24.

20. Statement by Richard Garwin, International Seminar on How to Avoid Nuclear War, Erice-Trapani, Sicily, August 20, 1983. Cited in Hugh DeWitt and Gerald Marsh, Stockpile Reliability and Nuclear Testing. Bulletin of the Atomic Scientists, April 1984: 40.

21. In the context of nuclear warhead design, the term "military characteristics" has a specific meaning: the requirements, as prepared by DOD in conjunction with DOE, for a new warhead in priority order. Examples include nuclear safety, warhead size and weight, yield, and conservative use of nuclear materials. Nuclear Weapon R&D and the Role of Nuclear Testing. Energy and Technology Review. September 1986. p. 7-8. This source discusses military characteristics and tradeoffs among them.

22. Foley, Sylvester, Responses to Questions for DOE Budget Hearing, in House Armed Services Committee, hearings, Department of Energy National Security Programs Authorization Act for Fiscal Years 1987 and 1988: 127-128.

23. This is a Department of Defense military characteristic (see note 19) for the W87, quoted in Nuclear Weapon R&D and the Role of Nuclear Testing, Energy and Technology Review, September 1986, p. 8.

24. Batzel, Roger, and Robert Thorn. Letter from Lawrence Livermore and Los Alamos National Laboratories Responding to Questions from Representative Henry J. Hyde. In U.S. Congress. House. Committee on Foreign Affairs and Its Subcommittee on Arms Control, International Security and Science. Proposals to Ban Nuclear Testing. Hearings and Markup on House Joint Resolution 3. 99th Congress, 1st session. Washington, U.S. GPO, 1985. p. 384-385.

25. LLNL Report: 48.

26. Nuclear Weapon R&D and the Role of Nuclear Testing. Energy and Technology Review, Sept. 1986: 18. Tritium has a half-life of 12.3 years.

27. Kidder, Ray. Letter to the editor, Physics Today, December 1983: 13. Emphasis in original.

28. DeWitt, Hugh, and Gerald Marsh,, Weapons Design Policy Impedes Test Ban. Bulletin of the Atomic Scientists, Nov. 1985: 10-13.

29. LLNL Report, p. 29.

30. Fetter, Steve. Toward a Comprehensive Test Ban. Cambridge, MA: Ballinger, in press.

31. Fetter, Stockpile Confidence Under a Nuclear Test Ban, p. 152.

32. Immele and Brown, A Commentary on "Stockpile Confidence During a Nuclear Test Ban," forthcoming, International Security. Draft, p. 12.

33. Rosengren, Some Little-Publicized Difficulties with a Nuclear Freeze, p. 19.

34. Letter from Herbert York to Jose Fulco and others, September 22, 1987.

35. Nuclear Weapon R&D and the Role of Nuclear Testing, Energy and Technology Review, September 1986: 8.

36. SAAC, p. 14-15.

37. Sources: (1) SAAC, p. 34. (2) The Nuclear Weapons Databook indicates (p. 63) that the B57 is a tactical bomb, so it is excluded here. (3) Regarding MX, U.S. Air Force, Air Force Systems Command, Headquarters Ballistic Missile Office, M-X Horizontal Shelter Weapon System: Baseline Configuration, Dec. 1980, p. II-71, indicates that the MX will accommodate 12 Mk 12A RVs or 11 Advanced Ballistic Reentry Vehicles (ABRVs). The Nuclear Weapons Databook indicates that the warhead for the Mk 12A is the W78 (p.75), and that the warhead for the ABRV, or Mk 21, is the W87 (p. 121). Regarding the Trident II, Rear Admiral Kenneth Malley, USN, Director, Strategic Systems Program, in a statement before the Subcommittee on Procurement and Military Nuclear Systems, House Armed Services Committee, March 9, 1988 (p. 3), indicates that Trident II "has the capability to carry a new, higher yield reentry body, the MK5. Alternatively, the MK4 reentry bodies, currently deployed on the C4 [Trident I missile], may also be carried." SAAC, p. 34, indicates that the W76 is the warhead for the Trident I. The W88 is the higher-yield warhead for the Trident II; see p. 63 of this study.

38. This topic is discussed in Fetter, The Comprehensive Test Ban, chapter 3, p. 18-21.

39. Barker, Robert. Debate on a Comprehensive Nuclear Weapons Test Ban: Con. Physics Today, August 1983: 27.

40. Soviet Union-United States Summit in Washington, DC: Joint State-ment. December 10, 1987. in U.S. Office of the Federal Register. Weekly Compilation of Presidential Documents. Dec. 14, 1987: 1494.

41. For warhead accuracy and yield, see U.S. Congress. Congressional Budget Office. Trident II Missiles: Capability, Costs, and Alternatives. by Jeffrey Merkley. Washington, U.S. GPO, 1986. p. 10.

42. Michael Gordon, U.S. Plans to Test Submarine Missile with 12 Warheads, New York Times, Oct. 7, 1987: 1, 11.

43. Twelve W76s could destroy almost as many silos hardened to resist 2,000 pounds per square inch (psi) of blast overpressure as eight W88s, but do progressively worse than eight W88s as silo hardness increases. The Soviet Union has reportedly increased the hardness of many silos above 2,000 psi. (Navy to Develop New Trident Warhead, Aviation Week and Space Technology, Jan. 17, 1983: 26. See also U.S. Department of Defense. Soviet Military Power, 1986. Washington, USGPO, 1986. p. 24.) It could harden silos further if it suspected that the United States had switched from W88s to W76s.

44. The W76 could destroy soft targets (with a hardness of 5 psi) to a range of two miles. For many soft targets, more yield is unnecessary. For large soft targets, more yield or more warheads would be needed.

45. Midgetman will use a variant of the W87 for MX.

46. Center for Defense Information. Simultaneous Test Ban: A Primer on Nuclear Explosions. Defense Monitor, vol. XIV, no. 5. Washington, 1985: 8.

47. U.S. Department of Energy. Briefing on Nuclear Testing. c. Aug. 1986. p. 9.

48. Scowcroft, Brent, John Deutch, and R. James Woolsey, Nukes: Continue the Tests. Washington Post, June 29, 1986: C-7.

49. This is particularly so since the cessation of atmospheric testing with the Limited Test Ban Treaty of 1963 virtually eliminated fallout from Soviet nuclear warheads, depriving the United States of a major source of information on their design. See Department of Energy, Briefing on Nuclear Testing, p. 9.

50. Meyer, Prof. Stephen. Department of Political Science, Massachusetts Institute of Technology, telephone discussion with Jonathan Medalia, September 19, 1986.

51. Collins, John, and Bernard Victory, U.S./Soviet Military Balance: Statistical Trends, 1977-1986 (As of January 1, 1987). U.S. Library of Congress. Congressional Research Service. Report 87-745-S. September 1, 1987, p. 17, 22.

52. See, for example, statement of Senator Edward Kennedy on amendment 2022 to S. 2355, FY89 national defense authorization bill, Congressional Record (daily ed.), May 12, 1988: S 5507; and Frank von Hippel, Harold Feiveson, and Christopher Paine, A Low-Threshold Nuclear Test Ban, International Security, Fall 1987: 147.

53. Corddry, Charles. 2 Former Officials Say U.S. Can Afford Nuclear-Test Cuts. Baltimore Sun, August 26, 1986: 1.

54. Responses by Donald Kerr to Additional Questions Submitted by Minority Members of the Subcommittee on Arms Control, International Security and Science, in U.S. Congress. House. Committee on Foreign Affairs. Subcommittee on Arms Control, International Security and Science, hearings and markup: Proposals to Ban Nuclear Testing. 99th Cong., 1st Sess. Washington, U.S. GPO, 1985: 379.

55. U.S. Department of Energy. Policy Paper 5: Nuclear Weapons Testing. Washington, U.S. GPO, January 1987: 37.

56. Kidder, Ray. Maintaining the U.S. Stockpile of Nuclear Weapons During a Low-Threshold or Comprehensive Test Ban. Lawrence Livermore National Laboratory, Oct. 1987, 33 p. UCRL-53820.

57. Kidder, Maintaining the U.S. Stockpile, p. 8.

58. Kidder, Maintaining the U.S. Stockpile, p. 17.

59. Kidder, Maintaining the U.S. Stockpile, p. 5; and U.S. Departments of Defense and Energy. The Effects of Nuclear Weapons, third ed. Compiled and edited by Samuel Glasstone and Philip Dolan. Washington: USGPO, 1977. p. 21.

60. Kidder, Maintaining the U.S. Stockpile, p. 8.

61. Kidder, Maintaining the U.S. Stockpile, p. 5.

62. Immele, John, and Paul Brown, A Commentary on "Stockpile Confidence During a Nuclear Test Ban," International Security, forthcoming; draft, p. 6.

63. Miller, George, and Paul Brown, Preliminary Comments on "Maintaining the U.S. Stockpile of Nuclear Weapons During a Low-Threshold or Comprehensive Test Ban," by Ray Kidder. December 10, 1987. Manuscript, p. 1.

64. Miller and Brown, Comments on "Maintaining the U.S. Stockpile...," p. 3.

65. Broad, William. U.S. Is Committed to Nuclear Tests. New York Times, Oct. 18, 1987: 1.

# 6

## Effects of More Restrictive Test Bans on Nuclear Effects Testing

*David Cheney and Robert Civiak*

Nuclear explosions produce a variety of effects, such as blast and several types of radiation, that can damage or destroy military systems. A program of both nuclear testing and extensive nonnuclear simulation of nuclear effects is used to make U.S. military systems that may need to function in a nuclear environment more resistant to these effects, and to make U.S. nuclear weapons more effective against Soviet targets. The effects testing program is used to characterize nuclear effects, to assess how damaging these effects are to military systems, and to validate that military systems meet their design criteria in surviving specified threat levels. More restrictive test ban treaties would, to varying degrees, restrict the nuclear testing that is part of this program. This chapter examines how these restrictions would affect U.S. interests.

Test ban opponents maintain that nuclear explosive effects testing is essential to assure confidence in the survivability of U.S. strategic weapon systems and command and control systems in order to preserve a credible deterrent. In addition, they maintain that effects testing is necessary to make new U.S. weapon systems, such as earth penetrating weapons, effective against Soviet targets.

Test ban supporters, in contrast, maintain that nuclear effects testing is one of many factors that affect the survivability of military systems and that its loss would at worst reduce confidence in the survivability of only new U.S. military systems at high threat levels, which would be offset by a test ban's constraints on the evolution of Soviet nuclear weapons. They note that a test ban would similarly restrain Soviet effects testing and that if deemed necessary, nuclear effects testing could continue under most threshold or quota treaties.

Both sides of the debate agree that maintaining confidence in the survivability of military systems is important. Both sides also agree that nuclear effects testing is useful in validating that systems can survive design threat levels and in improving the survivability and effectiveness of new U.S. weapons. The two sides disagree, however, about (1) whether constrained nuclear effects testing would have a significant impact on the survivability of military systems and (2) whether the primary effects of a more restrictive test ban--constraining nuclear warhead development--would outweigh the need for effects testing to improve the survivability of weapon systems.

## Technology Background

### *Effects of Nuclear Explosions*[1]

Nuclear explosions create a variety of potentially damaging effects, many of which are different from those produced by chemical explosions. The dominant effects depend on the environment in which the nuclear explosion occurs. <u>In the vacuum of space</u>, the main output of a nuclear explosion is X-rays radiated from weapon materials heated to temperatures in the tens of millions of degrees;[2] other outputs include neutrons, gamma rays, and radioactive debris. With no atmosphere, there is no blast effect, and radiation travels unimpeded outward from the explosion. This intense radiation can destroy or disable objects in space. Direct damage from X-rays is the primary threat generated by nuclear weapons against military assets in space. Testing for resistance to X-rays is a primary focus of underground nuclear effects testing.

When X-rays or gamma rays reach the atmosphere or objects in space they can cause secondary effects:

- Gamma rays that reach the Earth's atmosphere strip electrons from air molecules, creating charged particles that interact with the Earth's magnetic field to create a pulse of electromagnetic energy called "electromagnetic pulse" (EMP) that can damage electronic equipment. This type of EMP (which can also occur from explosions in the upper atmosphere) is called "high altitude EMP."

- X-rays that hit an object in space that contains electronic equipment (e.g., a satellite) can strip electrons from the object's surface material, creating an electrical surge in the equipment. This effect is called "system generated EMP."

The dominant effects of a nuclear explosion <u>in the atmosphere</u> are heat and a shock wave. That is because most of the radiation is quickly absorbed by the air, creating a fireball of hot gases and a shock wave. The shock wave and heat cause much of the damage from the explosion by knocking down structures and causing fires. Explosions in the atmosphere or near the Earth's surface also create a third type of EMP, called "source region EMP." Source region EMP occurs in a small area around the nuclear explosion and can damage electronic systems that are not destroyed by the blast, for example on missiles in hardened silos.

An explosion <u>near the Earth's surface or a shallow underground</u> explosion also creates craters and shock waves in the ground. Underground explosions, such as would occur from an earth penetrating warhead, produce much larger craters and stronger ground shock waves than surface explosions of similar yield. An explosion near the Earth's surface can also create a cloud of radioactive dust that can damage aircraft and cruise missiles which fly through it and carry lethal fallout downward.

## The Effects Testing Process

The Defense Nuclear Agency (DNA), an agency of the Department of Defense (DOD), conducts tests to characterize the effects of nuclear explosions, and to allow the Department of Energy (DOE) and the armed services to assess how damaging these effects are to materials, components, and U.S. and Soviet military systems, and to validate the resistance (or hardness) of U.S. systems to specified levels of nuclear effects. The most realistic tests for obtaining this information would be nuclear explosions in the environment in question, but treaty commitments and practical considerations limit testing to underground nuclear explosions and nonnuclear experiments in actual or simulated environments.

Determining the survivability of U.S. military systems is a complex process. During the design of a new system, the military establishes threat levels for various nuclear effects that the system will be designed to survive. In specifying the threat levels, the designers take into account the system's mission, actual and potential Soviet nuclear threats to the system, and estimates of the cost and performance penalties that would be incurred to survive various threat levels. No system could be designed to survive a direct nuclear hit, therefore design threat levels must be chosen at some lesser threat. For example, some U.S. reentry vehicles (RVs) are designed to be hardened so that no existing or anticipated Soviet nuclear ABM system could kill two RVs with one explosion.[3]

Next, the system is designed to withstand the prescribed threat, using the knowledge gained from previous testing experience. Typically, the knowledge is incorporated in analytical methods and computer models that predict how well materials, components, and systems will survive a given threat. These techniques have been refined to provide reasonable confidence in system survivability for current technologies against currently envisioned threats. This is a dynamic process; new materials, new types of electronic systems, and changes in the threat create a need for changes in the methods of hardening equipment, and these new methods need to be subjected to tests. In addition, testing is used to validate the adequacy of design.

Nonnuclear above-ground simulations are used for most testing of materials, devices, and systems because of the great expense and lengthy preparation time for underground nuclear tests. An underground nuclear effects test may cost more than $50 million and require several years of planning.[4] Using above-ground simulators, equipment can be tested, modified to correct deficiencies, and tested again repetitively at relatively low cost. Above-ground simulators typically simulate one effect at a time rather than all of the effects of a nuclear explosion simultaneously.

The final stage of the process is often an underground nuclear test to validate that the materials, devices, or systems will survive the design threat levels. In addition, nuclear tests are used to confirm or improve the basic techniques used to harden military equipment. A large test area may be exposed to radiation from a single nuclear explosion and in practice dozens of experiments are conducted during a single nuclear effects test. The experiments often include multiple tests of the same item to assess its ability to survive various threat levels.

Information from the Defense Nuclear Agency indicates that nuclear effects tests during development have revealed problems with one or more components in either the missile or RV that would have led to system failure at or near design threat levels for six of the eight U.S. strategic missile systems deployed during the past 25 years.[5] In three of those cases the system was modified and retested and survived the design threat level. In the other three cases either the design threat levels were modified or the operating characteristics of the system were changed to reduce the anticipated threat levels.[6]

As noted earlier, nuclear explosions in space and near the Earth's surface produce different effects and require different effects tests. The following two sections elaborate on the testing required for each environment.

## Testing for the Effects of Nuclear Explosions in Space

The direct effects of nuclear explosions in space are well understood. Other effects that result from the interaction of a nuclear explosion with the Earth's atmosphere and magnetic field, such as high altitude EMP, are not as well understood. Nuclear explosive tests in the upper atmosphere or in space would be useful in increasing knowledge of these effects, but they have been banned since 1963. Therefore, further understanding of these effects can only come from calculations and simulations.

As discussed above, assessing and validating the survivability of materials, components, and systems to specified threat levels is done through a combination of nonnuclear and nuclear tests.

*Nonnuclear Tests.* Radiation from nuclear explosions in space can be simulated with varying degrees of fidelity. Neutrons, X-rays, and gamma rays are produced by using lasers and particle accelerators to focus a tremendous amount of energy on a small target. When certain targets are energized they emit radiation similar to that produced by nuclear explosions (in some cases creating a small nuclear fusion reaction).[7] Different types of radiation can be produced by varying the targets and the input energy.

The U.S. capability to simulate the effects of nuclear explosions has improved greatly in recent years as more powerful generators, accelerators, and lasers have come on line. The improvements make it possible to generate higher intensities of radiation over larger areas. Nonnuclear simulation is now considered adequate for many but not all applications. All testing for high altitude EMP effects is done with non-nuclear simulators because underground nuclear explosions cannot generate this phenomenon.[8] The simulation of neutrons is adequate for most threats, although it may not be adequate for threats to future space-based SDI systems or threats to reentry vehicles from some possible Soviet ABM systems.[9] Gamma ray simulation does not perfectly imitate gamma rays from nuclear blasts, but is considered to be fairly good for hardening missiles or reentry vehicles.[10] It may not be adequate for testing some space-based SDI systems, which could be exposed to high threat levels and may use new materials whose resistance to nuclear effects are not well understood.[11]

Existing sources for stimulating the x-ray effects of nuclear explosion--the most lethal effect of nuclear explosions in space--are not fully adequate.[12] X-ray generators can simulate to some extent the effect of "hard" or "soft" X-rays[13] of sufficient energy and intensity to test small portions of equipment at desired threat levels, or to test full systems at lower threat levels. They are currently inadequate to test full systems at desired threat levels.

*Nuclear Tests.* All military systems that are based in or travel through space make use of data from underground nuclear tests to varying degrees. Those that are designed to withstand the highest radiation flux over the largest area are emphasized in the underground nuclear testing program. As described previously, nonnuclear simulators can generate realistic doses of radiation over small areas or smaller doses over larger areas, but cannot subject large systems to high radiation flux.

Table 9 is an unclassified summary of scheduled DNA underground nuclear effects tests. It provides a view of the priorities of the underground nuclear effects testing program (both those simulating explosions in space and explosions near the Earth's surface, which will be discussed later). The summary includes major experiments only; the actual tests will contain many other experiments not listed. After the list was provided, the U.S. testing program was set back by several months as a result of a labor strike at the Nevada Test Site and the schedule was revised substantially.

As can be seen from the table, the major objectives of the tests include determining the hardness of components of new missile systems, including the TRIDENT II and the small ICBM, and assessing the survivability of SDI systems.

# Table 9

## Defense Nuclear Agency Test Schedule
## (unclassified version, slightly condensed)

**MISSION CYBER**: 9/87.[14] The third of four tests in support of the TRIDENT II program; to include the MK-5 Re-entry Body Proof Test for final hardness verification and survivability assessment of the MK-5. Experiments are also planned to assess the vulnerabilities of the D-5 missile body electronic packages and the MK-6 guidance electronics. There will also be SDIO advanced development experiments associated with the Army's ERIS and HEDI ballistic missile defense programs, and SDI lethality and survivability experiments.

**MISTY ECHO**: 8/88. Cratering and ground shock event to develop energy coupling and cratering prediction codes to assess the vulnerability of strategic structures to nuclear attack. Also to help ensure the most effective allocation of strategic warheads for targeting purposes and to guide the development of future nuclear weapon systems including earth penetrating weapons.

**MINERAL QUARRY**: 3/89. Test in support of the Air Force Small ICBM. Includes experiments to establish degradation and failure threshold data for candidate subsystems and components for missile design. Other experiments include decoys and advanced reentry body components from the Air Force Ballistic Missile Office and tunnel hardening experiments from the Defense Nuclear Agency.

**DISCO ELM**: 9/89. Fourth and final test in support of the TRIDENT II program. The TRIDENT II Missile System Proof Test is planned to demonstrate systems survivability while the system is operating in a simulated boost phase flight profile. The final assessment of system capability will be verified by this system level test.

**HURON FOREST**: 9/90. Vertical Line of Sight event to support the SDI program (LTH-3) and will provide the critical lethality demonstration of large (scaled) structure response.

**HUNTERS TROPHY**: 3/91. Event in support of Strategic Defense Initiative Organization advanced development experiments associated with Army's ERIS and HEDI ballistic missile defense programs, SDIO survivability experiments (space components), decoys and advanced reentry body components from the Air Force Ballistic Missile Office.

**DIAMOND FORTUNE**: 9/91. Underground cavity event to assess the energy coupling and cratering effects of a shallow earth penetrating weapon. This event is an important element in the development of energy coupling and cratering prediction codes used to assess the vulnerability of strategic structures to nuclear attack. More importantly, it will assess the effectiveness of a shallow EPW against a high value target set, and guide development of future nuclear weapon systems.

**MINI URN**: 3/92. This event is in support of SDIO advanced development experiments associated with the Army's ballistic missile defense programs, SDIO survivability experiments, and decoys and advanced reentry body components from the Air Force Ballistic Missile Office.

**MIDNIGHT ZONE**: 9/92. A development test to evaluate key components and containment aspects of the MIDNIGHT AURA (SREMP) test. This test may also consist of a DOE proof test which includes the front-end and device output measurements.

**HYPER VIPER**: 3/93. This event is in support of advanced system design.

**DISTANT GALE**: 9/93. This event is in support of advanced system design.

**MIDNIGHT AURA**: 3/94. This event is an open air source region electromagnetic pulse (SREMP) test. The event is in direct support of vulnerability testing of Army tactical systems to SREMP on a large scale.

**MIGHTY WALE**: 9/94. This event is in support of advanced system design.

*Source: Defense Nuclear Agency, February 1987.*

## *Testing for the Effects of Nuclear Explosions Near the Earth's Surface*

Nuclear explosions near the Earth's surface or in the atmosphere are banned under the 1963 Limited Test Ban Treaty. To simulate these explosions, DNA uses above-ground nonnuclear tests and nuclear tests in underground cavities, which in effect create small earth surfaces. Nonnuclear techniques are best for testing structures and equipment against most heat and blast levels that may be generated by nuclear explosions. Underground nuclear tests are used to better understand the effects of nuclear explosions and to generate effects that occur very close to a nuclear explosion that cannot be well simulated by nonnuclear means.

*Nonnuclear testing.* Nonnuclear methods are the primary means of testing military devices for their resistance to heat, shock, EMP, and some types of radiation.[15] The heat and blast from an atmospheric nuclear explosion are fundamentally similar to those from conventional sources and are simulated in several ways. DNA has detonated the largest known chemical explosion--equivalent to eight kilotons of TNT--to simulate the heat and blast effects of an atmospheric nuclear explosion.[16] The blast, in June 1985, exposed replicas of jet aircraft, tanks, missile launchers, bunkers, radar systems, and submarines to a simulated nuclear airblast. Above-ground tests also provide information on dust generated from near surface explosions. That information is used in the design of airplanes and cruise missiles that might fly through such dust clouds.[17]

Other nonnuclear simulators include thermal flash facilities that can heat components to thousands of degrees Celsius,[18] and blast simulators that can simulate the effects of blast waves. France currently has the largest blast simulators. The United States, however, has conducted studies of a much larger facility that could reproduce the blast effects of a range of nuclear explosions.[19] Another nonnuclear technique is the high-explosive simulation technique (HEST), which is used to generate tens of thousands of pounds of overpressure to test models of missile silos.[20]

*Nuclear Tests*. Underground nuclear tests are used to develop data on the cratering and ground shock effects of surface, near surface, and shallow underground nuclear explosions. The Misty Echo test (see Table 9) is an example. The data produced by these tests has implications for the basing strategies of U.S. missiles, for the targeting of Soviet silos, for determining the optimum burst height for U.S. nuclear weapons, and for the feasibility of earth penetrating weapons.[21]

Underground cavity nuclear tests also generate information that is used in superhardening silos. This includes information on several nuclear effects that occur in the immediate vicinity of a nuclear explosion near the Earth's surface, including very high overpressure airblast, fireballs, and source region EMP.

## The Effects of a Comprehensive Test Ban Treaty on Effects Testing

Under a CTBT, the underground nuclear effects testing described above would be forgone. Without such tests, development of new military systems would be impaired by constraints on testing their effectiveness and by a decrease in confidence in their hardness, i.e. their ability to withstand nuclear effects. Pro- and anti-testing advocates dispute the extent to which confidence in hardness would decrease and the degree to which other means for improving system survivability can compensate for the loss of confidence in hardness.

While nonnuclear testing cannot as yet totally substitute for underground nuclear effects testing, as discussed above, significant improvements in nuclear effects simulation are foreseeable over time. Scientists believe that the next generation of inertial confinement fusion (ICF) machines,[22] with more powerful lasers or particle beam accelerators, will be able to produce all the radiation effects of a nuclear explosion at the same intensity currently tested for underground.[23] No such machine exists, however, and it would take at least eight years to design, build and test one.[24] Moreover, simulation using ICF may not reproduce the effects of nuclear explosions exactly. For example, the production of the required X-ray intensities from an ICF machine might be accompanied by different levels of neutrons and debris than would occur during nuclear attack against assets in space. Furthermore, improvements in the methods of modeling and simulating nuclear effects would be slowed under a CTBT by the inability to test new ideas against results from specially designed nuclear explosions. As a result, system designers' confidence in the hardness of new military systems would be less than if testing were to continue.

A loss of confidence in system hardness would act to decrease the confidence of military planners in the survivability of those systems, but other measures could compensate for the loss. Hardness could be maintained by designing systems more conservatively, for example, by adding shielding, using more resistant materials, or using electronic circuitry that is less susceptible to damage. These measures might reduce system performance or increase cost, however, and without underground nuclear tests to validate hardness, designers' confidence in the hardness of even conservative designs might not be as high as if testing were continued.

Survivability depends on many factors in addition to hardness. System designers could compensate for reduced confidence in hardness by improving other survivability factors. Options to improve survivability include: (1) using evasive measures such as stealth, decoys, or maneuvers to avoid attack; (2) altering system basing modes, such as by placing satellites further from one another or in higher orbits; (3) increasing redundancy in the system, for example by deploying more missiles or satellites; (4) altering operational strategies to avoid attack, such as adopting a launch on warning strategy or rapidly deploying lightweight satellites just before they are needed; or (5) using active defensive measures such as shooting at attacking missiles. These alternatives all have costs and none may be practical for a particular system. The point in listing them is to demonstrate that hardness is only one means of improving survivability and a detailed study of a particular system and its mission has to be made to determine whether there would be an unacceptable loss in system survivability (or even an unacceptable loss in confidence in system survivability) with a loss of effects testing.

It should also be noted that survivability of any weapons system can never be 100 percent assured because of uncertainty in the actual nuclear effects that a military system will face, including the possibility of a direct hit.

In addition to constraining effects testing for improving the hardness of military systems, a CTBT would hamper effects testing used to improve the effectiveness of new weapons, such as cratering tests for improving earth penetrating warheads. The result would be an additional constraint on the development of new nuclear weapons in both the United States and the Soviet Union and would enhance the direct constraints of a CTBT on new weapons development discussed in chapter 4.

The survivability of existing military systems would be affected by the loss of underground nuclear effects testing under a CTBT to a lesser degree than new systems. Two counteracting influences would affect the survivability of existing systems. On the one hand, effects testing useful in improving the survivability of existing systems, such as for further hardening underground command centers or missile silos, would be forgone. On the other hand, development of Soviet weapons that further threaten those systems would be constrained. It is not clear if the net effect of those competing influences would enhance or diminish the survivability of existing military systems.

### Missile Systems

The only weapons currently beyond phase 2a of development (design definition and cost study) that travel through space and thus might be affected by a halt in nuclear explosive effects tests are the Trident II missile (D-5) and the small ICBM (SICBM or Midgetman). The MK-5 reentry vehicle for the Trident II underwent a nuclear explosive effects test during the Mission Cyber test in December 1987 (see Table 9) as a final proof test to verify its hardness and assess its survivability. The electronics and guidance systems of the D-5 missile body were tested for hardness at the same time. A full system test of the Trident II in a simulated boost phase is scheduled for September 1989.

Unless the Mission Cyber test uncovers a significant problem requiring another test to correct, the only potential impact of a CTBT on effects testing for the Trident II system would be the loss of the proof test of the system in the boost phase. Apparently the September 1989 test will assess the vulnerabilities of the Trident II to future strategic defensive systems since the boost phase normally would take place well beyond the range of current Soviet ABM systems and is not susceptible to targeting by first-strike offensive weapons. Therefore, it seems that Trident II deployment would be relatively unaffected by a CTBT. If the Soviet Union deploys nuclear systems that threaten the Trident II in its boost phase in the future, the survivability of the Trident II would be more difficult to assess in the absence of the final test than if it were conducted. A CTBT would make such Soviet deployment unlikely.

The situation regarding effects testing of the SICBM is similar. Modified versions of the warhead (W87) and RV (MK-21) used on the MX missile have been selected for the SICBM.[25] The W87 and MK-21 have been thoroughly tested against nuclear effects. The DNA has scheduled nuclear explosive effects testing of subsystems and components of candidate missile designs for March 1989. It is likely that further effects tests of the missile systems would be scheduled as the design proceeds. Therefore, it appears that a CTBT would not affect the confidence of military planners in the ability of the SICBM warhead and reentry vehicle to withstand nuclear effects, but it could reduce confidence in the ability of the missile system to withstand potential future nuclear attack. However, Midgetman survivability will rely predominantly on its mobility, and it is not clear how additional uncertainty in its ability to withstand nuclear effects would alter deployment plans.

### Satellites

As with other systems, satellites are not hardened to survive direct attacks, but are hardened and deployed so that one nuclear antisatellite weapon cannot kill two satellites. Hardened satellites can survive a 1-megaton explosion 100 kilometers away; unhardened satellites can be destroyed by high yield nuclear explosions several thousand kilometers away. Satellites used for warning and communications are in high earth orbits--thousands of kilometers high--and, if hardened, are largely invulnerable to being destroyed by nuclear explosions that are not directed at them. Satellites in low orbits, such as some reconnaissance satellites, are more vulnerable. They are closer to each other and to the earth, and thus can be more easily reached by anti-satellite weapons and are more likely to be damaged by nearby nuclear blasts.[26]

Because satellites are typically far apart, particularly the communications, navigation, and early warning satellites that are in high orbits, they do not have to be designed to withstand as intense radiation as RVs. It appears that satellite survivability would suffer little under a CTBT. Nonnuclear simulation can provide sufficient radiation flux for their effects testing, except for complete system tests of large satellites at high threat levels. Complete satellites are also difficult to test in underground nuclear tests and none have been so tested, although satellite subsystems and two satellite mockups have been tested.[27] The major purpose of the mockup tests was to verify existing methods of hardening satellites. The methods were found to be generally effective,[28] but in both tests minor malfunctions occurred that leave some uncertainty about whether the mockup tests provide sufficient confidence in the methods of hardening satellites[29].

On the other hand, if it is determined that satellites must withstand more intense radiation than they currently can resist, for example to survive attack by a potential future nuclear antisatellite (ASAT) weapon, more nuclear effects testing of satellites might be needed. A CTBT, however, would limit new threats to satellites. For example, development of an X-ray laser, which in theory could be a potent antisatellite weapon, would be precluded under a CTBT.

### Strategic Defense Initiative

A definitive SDI architecture has not yet been established, but any space-based ballistic missile defense (BMD) is likely to contain sensors and communication satellites in high Earth orbit as well as battle stations in low Earth orbit. Space-based BMD components can be expected to be targets in a nuclear attack, and the system would have to be survivable for it to provide an effective defense. Several nuclear effects tests listed in Table 9 are scheduled to assess the survivability of SDI systems against possible Soviet nuclear threats.

A detailed discussion of SDI survivability is beyond the scope of this report, but survivability will depend upon evasive and defensive measures as well as hardening.[30] The threat levels to which SDI systems might need to be hardened will depend upon the success of the other measures, but it is possible that space-based SDI systems will have requirements for radiation hardening at threat levels considerably above those of current satellites. In that case, existing nonnuclear techniques for simulating nuclear effects might be insufficient to validate the hardening of SDI systems. The absence of nuclear explosive effects tests under a CTBT could make the difficult task of development of space-based strategic defensive systems even harder.

The difficulty in validating the survivability of space-based BMD components under a CTBT could be compounded by the SDI program's plans to use many new materials that have not been subjected to effects tests to the same extent as the materials in warheads and missiles.

An SDI system would require good communication between various land- and space-based components of the system during a nuclear exchange. Some analysts feel that this would be difficult to ensure without nuclear testing in the atmosphere and in space to examine how EMP and other phenomena will affect communications.[31] The United States conducted only one nuclear test in the upper atmosphere to examine EMP before the Limited Test Ban Treaty prohibited such tests. DNA officials believe, however, that the effects of nuclear explosions in space and in the upper atmosphere can be sufficiently well understood to be accommodated without conducting nuclear tests in those environments.[32]

## Strategic Consequences of Limiting Nuclear Effects Testing

### Consequences of a CTBT

Leaders of both the United States and the Soviet Union might have less confidence in the survivability of <u>new</u> strategic weapon systems (e.g., ballistic missiles) introduced during a CTBT because of a lack of nuclear effects testing. As discussed in Chapter 4, a CTBT would constrain new weapon development by stopping nuclear tests of warheads. The absence of nuclear effects testing would enhance that constraint.

If new weapons were deployed during a CTBT, the absence of nuclear effects testing might have a greater effect on the United States than the Soviet Union. Soviet missiles have large throw-weights, allowing for relatively large and heavy warheads and RVs that could carry more shielding than U.S. warheads and RVs.[33] The smaller throw-weight of U.S. ballistic missiles places a premium on using small, lightweight warheads that sharply limits the amount of radiation shielding that can be used. In this circumstance, realistic effects tests would be more important for the United States for optimizing shielding.

This asymmetry could be mitigated in several ways. Improvements in nonnuclear simulation could limit any loss of confidence in the survivability of new weapon systems. The consequences for survivability of strategic systems resulting from an inability to conduct effects tests could be nearly eliminated if these weapons were not exposed to nuclear effects. They would not be if the ABM Treaty remains in force, neither side deploys strategic defenses, and warheads are targeted so that early-arriving warheads do not interfere with following warheads. In any event, a CTBT would not affect confidence in the survivability of existing weapons.

With regard to satellites, even a balanced U.S. and Soviet loss of confidence in the survivability of communications satellites under a CTBT could have adverse strategic implications. Reduced confidence in these systems could conceivably lead one side to strike first in a crisis for fear of losing command and control capability. As discussed above, however, nonnuclear simulations appear adequate to harden satellites to the level necessary to survive existing threats. It thus appears that satellite survivability for existing threats would suffer little, if at all, from a CTBT, and evasive and defensive techniques can enhance satellite survivability against current or future threats without improved hardening.

The last area of concern is the potential for reduced confidence in the survivability of hardened underground structures under a CTBT. As with communication satellites, reduced confidence in the survivability of underground command structures or fixed missile silos could be destabilizing by adding to the pressure for early launch of nuclear weapons during a crisis. The absence of underground nuclear effects testing under a CTBT would constrain efforts to improve the hardening of these underground structures. On the other hand, development of new weapons that threaten underground structures, such as earth penetrating warheads, would also be constrained.

### Consequences of Lower Thresholds

The United States has not conducted a nuclear effects test with a source having a yield above 20 kt since 1966[34] and no plans for such a test are known. Therefore, it does not appear that a 20-kt LYTT would have any effect on the U.S. capability to conduct effects tests.

The situation with regard to a 1-kt LYTT is less clear. In principle, most of the knowledge obtained from underground tests can be obtained from explosions with yields less than 1 kt.[35] In particular, it might be possible to perform all current radiation hardening experiments at that yield. However, the radiation from a 1-kt source in general has relatively fewer high energy X-rays than radiation from a strategic nuclear weapon and the possibility of making an adequate 1-kt source has not been demonstrated.[36]  Therefore, it is not yet clear whether a 1-kt LYTT would be significantly different from a CTBT with regard to effects testing.

## *Consequences of a Quota Treaty*

On average, the United States conducts about two underground nuclear effects tests per year.  Therefore, if given priority over other testing requirements, nuclear effects testing could continue unabated under a quota of as few as two tests per year.  However, it is impossible to say what the priorities for U.S. nuclear testing would be under a Quota Treaty.

A Quota Treaty appeals to those who believe that new nuclear weapons development should be constrained, but that there could be adverse consequences from constraining effects testing of satellites or hardened structures and perhaps a loss of confidence in the existing nuclear weapons stockpile (see Chapter 5) from a halt in testing.  If that view were adopted as official U.S. policy, there might be little falloff in underground nuclear effects testing under a Quota Treaty.  There would be little, if any, strategic consequences of a small falloff in effects testing, because development of U.S. and Soviet warheads would be slowed under a Quota Treaty reducing the number of new U.S. warheads on which effects tests must be conducted and reducing the number of new Soviet warheads that compel effects testing on U.S. systems.  In addition, improvements in nonnuclear simulation could further reduce the requirements for nuclear effects testing.

On the other hand, if other testing requirements are given priority under a Quota Treaty and no underground nuclear effects are conducted, the strategic consequences in this area would be the same as under a CTBT.

# Notes

1. The major source for this section is Glasstone, Samuel, and Philip J. Dolan, ed. The Effects of Nuclear Weapons (unclassified edition). United States Department of Defense and Department of Energy. Washington, Govt. Print. Off., 1977. 653 p.

2. In the same way that metal becomes first "red hot" and then "white hot" as it is heated, nuclear explosions heat materials to temperatures such that they radiate energy as X-rays.

3. Dr. Cyrus P. (Skip) Knowles, R & D Associates and former head of the Defense Nuclear Agency's nuclear testing program. Private conversation, Nov. 24, 1987.

4. Attachments to memorandum of February 18, 1987, for Robert Civiak from the Defense Nuclear Agency.

5. Defense Nuclear Agency. Nuclear Effects Testing. Briefing book. (1987.)

6. Ibid.

7. Craxton, R. Stephen, Robert L. McCrory, and John M. Soures. Progress in Laser Fusion. Scientific American, vol. 255, Aug. 1986: 68-79.

8. Defense Nuclear Agency. Nuclear Effects Testing. Briefing book. (1987). Also, Kennedy, David M. Simulating the Dreaded Day. Technology Review, January 1985: 70-75.

9. James Powell, Sandia National Laboratory, telephone conversation with David Cheney, November 19, 1987.

10. Ibid. Also, Horgan, John. Underground Nuclear Weapons Testing. IEEE Spectrum, April 1986: 41.

11. James Powell, Sandia National Laboratory, telephone conversation with David Cheney, November 19, 1987.

12. Defense Nuclear Agency. Nuclear Effects Testing. Briefing book. (1987). DNA frequently emphasizes that inadequacies in x-ray simulation are a major reason for continuing to test underground. Also, James Powell, Sandia National Laboratory, telephone conversation with David Cheney, November 19, 1987.

13. "Hard" X-rays refer to X-rays at the high energy, high frequency end of the X-ray spectrum; "soft" X-rays are those at the low energy, low frequency end of the X-ray spectrum. Soft X-rays are more likely to damage the surface of an object; hard X-rays can penetrate surface materials and damage internal components.

14. This test was conducted on December 2, 1987.

15. Horgan, John. Underground Nuclear Weapons Testing. IEEE Spectrum. April 1986: 32-43.

16. Flory, Robert A. Expanded Simulation Techniques: Direct Course--a 1-kt height of burst [i.e., airburst] nuclear blast simulation; Minor Scale--an 8-kt surface nuclear blast simulation. Alexandria, VA, Washington Research Center. August 1986. 19 p. Available from NTIS, report no. AD-P-005361/1/XAB.

17. Statement of Lieutenant General John L. Pickitt, USAF, Director, Defense Nuclear Agency, before the R&D Subcommittee, Armed Services

Committee, House of Representative, Fiscal Year 1987, February 19, 1986. Unpublished testimony provided by the Defense Nuclear Agency, February 18, 1987.

18. Taylor, William F. Thermal Radiation Source Test Facility. Washington, Defense Nuclear Agency. January 1984. 32 p. Available from NTIS, #AD-A146561.

19. Opaklka, Klaus O. Large Blast-Wave Simulators (LBS) with Cold-Gas Drivers: Computational Design Studies. Aberdeen, Maryland, U.S. Army Ballistic Research Laboratory. March, 1987. 42 p. Available from NTIS, no. AD-A181 400.

20. Sanai, M. and Colton, J.D. Simulation Development and Silo Test Program (STP). Defense Nuclear Agency. March 31, 1984. 108 p. Available from NTIS, no. AD-A163 274. Also Kennedy, David M. Simulating the Dreaded Day. Technology Review, January 1985: 70-75.

21. Lawrence Livermore National Laboratory. Earth-Penetrating Weapons. Energy and Technology Review. June–July 1986: 4-5.

22. Inertial confinement fusion (ICF) is a technique in which a tiny pellet of fusion fuel is compressed and heated by bombarding it with powerful lasers or particle beams. The resultant pellet behavior emulates processes that occur in a thermonuclear explosion, including nuclear fusion. ICF can be useful in designing thermonuclear weapons as well as in producing the effects of a thermonuclear explosion on a small scale.

23. Eric Storm of the NOVA facility at Lawrence Livermore National Laboratory, personal interview with David Cheney, October 23, 1987. Also, James Powell, Sandia National Laboratory, telephone interview with David Cheney, November 19, 1987.

24. Kirk, Don, manager of the Sandia National Laboratory's Fusion Research Department, telephone conversation with David Cheney, November 16, 1987.

25. Statement of James Culpepper, Acting Assistant Secretary for Military Application, Department of Energy. In U.S. Congress. House. Committee on Appropriations. Subcommittee on Energy and Water Development. Energy and Water Development Appropriations for 1988. Hearings, 100th Cong., 1st Sess. Washington, Govt. Print. Off., 1987. Part 6, p. 829.

26. May, Michael M. Safeguarding Our Military Space Systems. Science, vol. 232, April 18, 1986: 336-340.

27. Defense Nuclear Agency. Nuclear Effects Testing. Briefing book. (1987.)

28. Scott, William B. Radiation Hardening Found Effective. Aviation Week and Space Technology, March 15, 1982: 71-74. Also, Shelton, Dr. Frank H. Satellite System Survivability. Proceedings of the 1983 Symposium on Military Space Communications and Operations, August 2,3, and 4, USAF Academy Colorado. The Electrical Engineering Department of the United States Air Force Academy. p. 29-31.

29. In both tests invalid data was found stored, after the test, in the attitude control subsystem. It is not clear if the malfunctions reveal a real problem or if they are artifacts of the testing environment that would not

occur in space. The issue is still under investigation. Defense Nuclear Agency. Nuclear Effects Testing. Briefing book. (1987)

30. See, for example, U.S. Department of Defense. Strategic Defense Initiative Organization. Report to the Congress on the Strategic Defense Initiative. (Washington) April 1987. Chapter VI.E.

31. Taylor, Theodore B. Nuclear Testing is a Pandora's Box. FAS Public Interest Report, v. 39, December 1986: 9.

32. Defense Nuclear Agency. Comments on CRS draft report. October 7, 1987.

33. There is no unclassified indication of whether Soviet warheads in fact are more resistant to nuclear effects than are U.S. warheads.

34. U.S. Department of Energy. Nevada Operations Office. Announced United States Nuclear Tests. July 1945 through December 1986. NVO-209 (rev. 7). January 1987. 66 p.

35. Feiveson, Harold A., et al. A Low-Threshold Test Ban Is Feasible; and Miller, George H., et al. Facing Nuclear Reality. Policy Forum. Science, vol. 238, October 23, 1987: 455-464.

36. Feiveson, Harold A., et al. A Low-Threshold Test Ban Is Feasible; and Miller, George H., et al. Facing Nuclear Reality. Policy Forum. Science, vol. 238, October 23, 1987: 455-464.

# 7

---

# Verification Issues

*David Cheney*

## Introduction

The capability to verify compliance with an arms control treaty is a necessary but not sufficient condition for the treaty to attract broad support in the United States. Test ban opponents can probably defeat a potential treaty if they convince people that verification is inadequate. Convincing people that verification is adequate removes an argument against a treaty, but does not necessarily enhance support for a treaty. One reason the United States puts a high premium on verification is that weaknesses in verification are potentially asymmetrical; it is generally accepted that because the Soviet Union is a less open society than the United States, it would be easier for the Soviet Union to test clandestinely.

It is important to distinguish between the terms "monitoring" and "verification." Monitoring is the technical function of observing activities; it goes on independent of treaties. Verification is the process of verifying--establishing the truth of--compliance or noncompliance with a treaty (or observance of a moratorium). Verification is the broader function: decisions about compliance or noncompliance are based in important part on information derived from monitoring, but also involve more subjective judgments concerning, for example, the precise interpretation of a treaty and Soviet intent. There is more to verification than monitoring.

Confidence in monitoring is never perfect; a country can never be sure of identifying all possible treaty violations. The imperfection leads to two major questions about verification capabilities. One is "How well can we determine compliance or noncompliance with a treaty?" This is largely a question about our technical capabilities to monitor Soviet activities. In theory, this question can be answered in terms of the probability of detecting certain types of violations. In practice, however, monitoring capabilities are imperfectly known and are thus debated in both technical and political arenas. The second question is "Are these capabilities acceptable?" This is a policy judgment that depends in part on the monitoring capabilities but also on the perceived military significance of possible violations, on the perceived benefits of a treaty, on one's views of Soviet intentions, and on other considerations. Agreement on the capability to monitor Soviet actions does not necessarily lead to agreement on whether the capabilities are acceptable.

Test ban opponents tend to demand stringent monitoring provisions and require high levels of confidence in the U.S. ability to detect treaty violations in order to prevent the Soviets from cheating or taking advantage of ambiguities. Test ban supporters tend to accept less stringent verification provisions in part because they see modest military consequences if the Soviets go slightly beyond agreed upon limits (and thus see little reason for the Soviets to risk cheating) and in part because they believe that the potential benefits of treaties outweigh the risks of Soviet noncompliance. People who hold these views are prepared to trade off some ability to monitor Soviet activities in return for more restrictive agreements.

Two standards of verification reflect these different views. The Nixon, Ford, and Carter Administrations used the term "adequate verification," which was defined by the Carter Administration as the ability to detect cheating of significance to the strategic balance in time to take appropriate action.[1] It is a somewhat ambiguous standard--people who disagree about what constitutes changes in the strategic balance are likely to disagree about the monitoring capabilities required. The

Reagan Administration viewed that standard as too loose to protect U.S. interests, and demands a stricter standard, which it terms "effective verification":

> Effective verification of compliance with the provisions of any arms control agreement usually requires that a single meaningful violation must be detected. In the case of nuclear testing limitations, it is generally agreed that a single test--even at low yield--in violation of a limitation can be critical to the side conducting the test if the test were designed to restore confidence in a weapon whose performance had become suspect, to proof test a new warhead, or to ensure the survivability of vital national security assets.[2]

This strict standard of verification requires more capable and usually more intrusive monitoring systems than does adequate verification.

Every test ban treaty has different verification requirements and raises different verification issues. With respect to the Threshold Test Ban Treaty (TTBT) and the Peaceful Nuclear Explosions Treaty (PNET) which are before the Senate for ratification consideration, the main verification issues are:

- Has the Soviet Union complied with the 150-kt limit of the TTBT?
- Do the verification provisions in the treaties provide for acceptable verification?
- If verification needs to be improved before the treaties are ratified, what verification provisions should be added?

Each of these issues is contentious. The Reagan Administration asserted that several Soviet tests likely have violated the TTBT. The Soviet Union and some American scientists dispute these charges. The Reagan Administration also argued that the methods the treaty provides for estimating the yields of Soviet explosions allow too much room for potential Soviet cheating. At issue is the quality of the methods and how good is good enough. The Administration has made the use of a particular on-site method (CORRTEX) of estimating yield a prerequisite for supporting the treaties.

There is debate about whether CORRTEX is necessary to improve verification.

For future treaties, for which the verification provisions have not yet been determined, the verification issues are somewhat different. For a comprehensive test ban treaty (CTBT), there is general agreement that for any given verification regime, clandestine explosions below some yield cannot be reliably identified. The main issues are:

- How fully can compliance with a CTBT be verified under various practical verification regimes?
- How significant is possible cheating?
- How acceptable are these risks in comparison to the possible benefits of the treaty?

The Reagan Administration believed that a CTBT is not effectively verifiable. Some test ban supporters agree, and support a 1-kt threshold treaty instead of a CTBT. Others assert that a CTBT would be adequately verifiable in the sense that clandestine low-yield testing under a CTBT would not affect the strategic balance.

A major reason for interest in a low-yield threshold treaty is concern about weaknesses in verifying a CTBT. With respect to a low-yield threshold treaty, the verification issues are:

- At what threshold and with what seismic monitoring system can explosions be detected and identified anywhere in the Soviet Union with acceptable confidence?
- What verification provisions would be required to estimate yields of explosions at the test site with sufficient accuracy to make the treaty acceptable?
- Can such verification regimes be negotiated?

If a quota treaty is deemed desirable to allow a limited amount of testing, verification issues associated with such a treaty include:

- Can one design a treaty with verification measures to alleviate concerns about testing several nuclear devices in one test?
- How should a treaty and verification measures be designed to alleviate concerns about conducting many tests below the detection threshold? (For example, should there be a low threshold below which unlimited testing would be permitted? If so, how should compliance with this threshold be verified?)

Verification issues surrounding low threshold test bans and quota treaties have not been widely discussed in the open literature of the technical community or in policy statements by the Administration.

## Technical Background

The method used for monitoring nuclear tests depends on the test environment. Underground tests are monitored largely by seismic methods. Underwater tests are monitored by acoustic methods using underwater listening devices employed primarily for antisubmarine warfare. Space tests are monitored primarily by remote sensing by satellites. Atmospheric tests are monitored by remote sensing and by analysis of fallout gathered by aircraft, ships, and ground stations. Reconnaissance satellites, electronic eavesdropping, and human sources such as defectors all provide additional information useful in detecting and identifying noncompliance. The following sections describe seismic and other monitoring methods.

### *Seismic Monitoring*[3]

Underground explosions create waves of vibrations that travel through and over the earth. These waves, called seismic waves, are similar to those created by earthquakes. They can be picked up by sensitive instruments, called seismometers, at great distances from the explosion. Monitoring these waves is the main method of monitoring underground nuclear tests.

Seismic monitoring has three main tasks. The first is to detect and locate the origin of seismic waves that are caused by earthquakes, nuclear explosions, or chemical explosions. The second task is to identify the origin of the seismic signal as an explosion or an earthquake. The third task, applicable to threshold treaties, is to estimate the yield of a nuclear explosion at a test site from the size of the seismic signal.

Seismic waves are analogous to sound waves. Both can be of high or low frequency (pitch), and can consist of recognizable waves and background noise (the waves from many sources jumbled together). Background seismic noise comes from such sources as wind, the crash of waves against a shore, and the movement of heavy trucks. Seismically "quiet" sites--the best places to listen for explosions--are removed from these sources of noise.

There are several basic types of seismic waves. Some travel through the body of the earth while others travel in the earth's crust or over the surface of the earth. This diversity of seismic waves is advantageous for monitoring explosions: each type of seismic wave carries different information that can help in detecting, identifying, and estimating the yields of nuclear explosions. Indeed, the greatest recent gains in seismic monitoring of underground explosions have come not from progress in seismometers but from a better understanding of how to process the data they produce and especially how to relate the differing pieces of data.

To detect an explosion, its seismic signal must have different characteristics than the background noise. Identifying a seismic signal as an explosion primarily involves the use of one of several discriminants to determine that the signal does not result from an earthquake. One discriminant is location. If a seismic event can be determined to have occurred deeper than current drilling capabilities, it can be determined to be an earthquake; if it is located in an area where earthquakes are rare, it will be suspected to be an explosion. Other discriminants make use of the distinctive seismic signatures of earthquakes and explosions. In the same way that different musical instruments create different sounds, earthquakes and explosions create different and recognizable patterns of seismic waves. According to theory, explosions create simple, relatively high frequency (i.e. high pitch)[4] seismic sounds; earthquakes in general create lower-pitched and more complex seismic sounds. In practice, however, complexities of the earth make it difficult to distinguish between some earthquakes and some explosions. About 15 or 20 methods, of varying utility, have been developed to distinguish explosions from earthquakes.[5] No single discriminant has been shown to work in all situations, but if enough information is available the combination of several methods can identify nearly all seismic events.

Figure 5, which shows a small explosion at the Soviet test site set off shortly after a large earthquake, illustrates some of the concepts relevant to detecting and identifying seismic events. Figure 5a is the record of these events monitored at the relatively low frequencies where the seismic waves of earthquakes should predominate. The figure illustrates how the seismic event signal must be larger than background noise for detection. The vibrations that precede the earthquake are background noise. In this case, the signal from the earthquake is much larger than the background noise, and is easily detected. If the signal from an earthquake was much smaller, perhaps a twentieth of the size, it would have been much more difficult to distinguish from the background noise at this station.

Figure 5b is the record of the same events as monitored in (a), except that it has been filtered to "listen" to higher frequency sounds.[6] This decreases the size of the signal from the earthquake and accentuates the signal from the explosion. Even though it is a small explosion, its signal stands out from that of the large earthquake. This illustrates how one of the differences between earthquakes and explosions--that explosions have relatively more of their energy at high frequencies than do earthquakes--could be used to help distinguish explosions from earthquakes.[7]

The third task of a monitoring system, necessary for verifying compliance with threshold treaties, is to estimate the yield of an explosion once it has been detected and identified.

## Figure 5

### Recording of a Small Explosion During an Earthquake

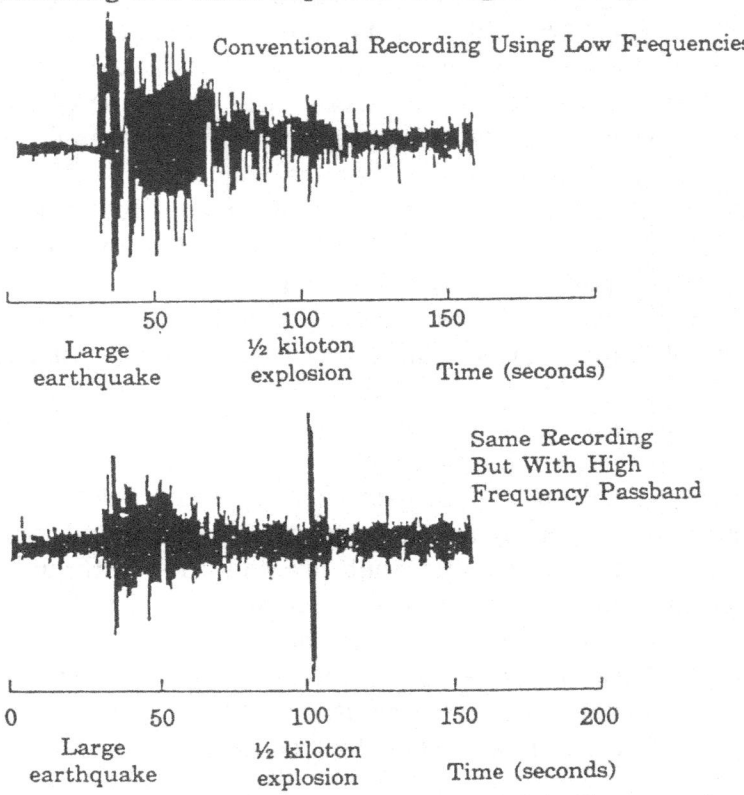

Conventional Recording Using Low Frequencies

|        50        |        100        |       150       |

Large          ½ kiloton
earthquake      explosion        Time (seconds)

Same Recording
But With High
Frequency Passband

| 0 | 50 | 100 | 150 | 200 |

Large          ½ kiloton
earthquake      explosion        Time (seconds)

Both the upper and lower seismograms were recorded in Norway and cover the same period of time. The upper seismogram is the conventional recording of low-frequency seismic waves. Both it and the lower recording show, at the time of 30 seconds, the arrival of waves from a large earthquake that occurred in the eastern part of the Soviet Union. About one minute after the earthquake, at a time of 100 seconds, the Soviet Union conducted a very small (about ½ kt) underground nuclear explosion at their Kazakhstan test site. With the standard filter (upper seismogram), the signal of the explosion appears hidden by the earthquake. Using a passband filter for higher frequency seismic waves (lower seismogram), the explosion is revealed.

Source: Semiannual Technical Summary for Norwegian Seismic Array for 1964, Royal Norwegian Council for Scientific and Industrial Research, Scientific Report No. 1-84/8S. Reprinted from U.S. Congress. Office of Technology Assessment. Seismic Verification of Nuclear Testing Treaties (1989), p. 12.

In the same way that the loudness of a sound provides information about the intensity or energy of its source, seismologists estimate yields of explosions by their seismic magnitude.[8] Figure 6 illustrates the basic method and some of the problems associated with it. It is an unclassified plot of seismic magnitudes of explosions at the Nevada Test Site versus the yield of the explosion. To estimate the yield of an explosion, one determines what yield of explosion (on the bottom axis) corresponds to the observed magnitude of the seismic signal (on the left hand axis). Note, however, that the relationship between magnitude and yield is different in different types of rock--explosions set off in dry tuff or alluvium (soft or porous rocks) create smaller seismic signals than do explosions set off in wet or hard rocks. Note also that even for explosions in the same type of rock, there is some scatter in the data; the points do not all fall exactly on the line. This scatter results in some uncertainty when estimating yields from magnitudes.

An additional complication (not shown in the figure) is that due to differences in the earth underneath the Soviet Union and the United States, the relationship between magnitude and yield is different for explosions in the Soviet Union than for explosions in Nevada. The relationship between magnitude and yield is well known for tests at the Nevada Test Site but must be adjusted for the Soviet geology to estimate the yields of Soviet explosions from the magnitude of their seismic waves. Seismic waves from Soviet explosions are believed to be larger than the seismic waves from equivalent explosions in Nevada.[9] Many studies have been conducted to attempt to determine how large an adjustment is needed to apply methods based on U.S. tests to Soviet tests.[10] The precise adjustment, which will be discussed later, is critical in determining the size of Soviet explosions.

***Potential Ways to Cheat.*** Several strategies have been devised to attempt to confuse monitoring methods and thereby prevent the other side from determining noncompliance with a treaty. These strategies need to be considered when determining how well compliance with treaties can be verified.

**Figure 6**

**Magnitude (Body Wave) Versus Yield Relationship
for Nuclear Explosions at the Nevada Test Site**

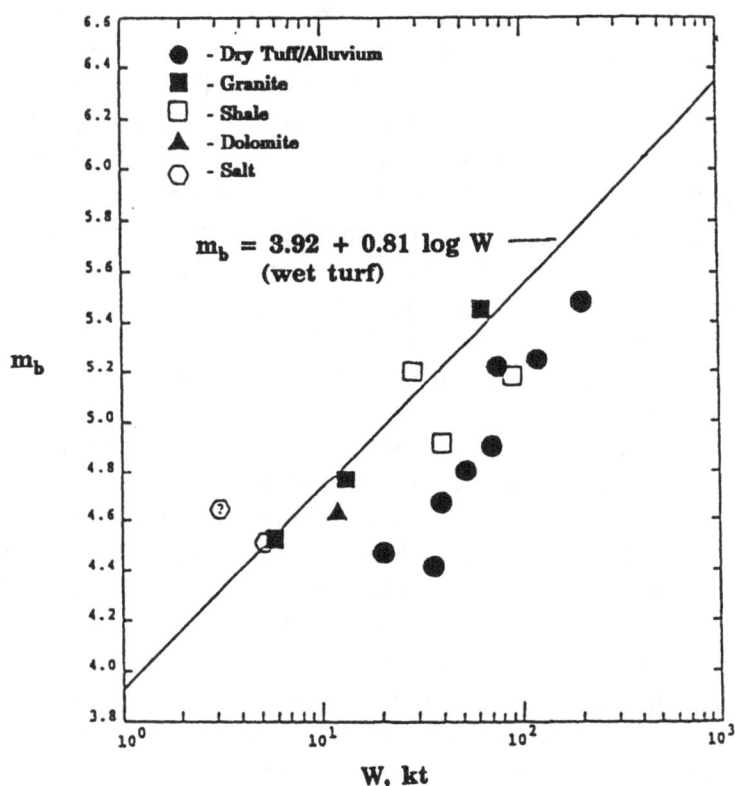

The line is drawn for explosion in hard rocks. Note that explosions in dry tuff or alluvium (solid circles) produce much smaller seismic waves than explosions of the same yield in hard rocks.

Source: John R. Murphy. "P Wave Coupling of Underground Explosions in Various Geologic Media." In: E.S. Husebye and S. Mykkeltveit, ed. Identification of Seismic Sources: Earthquake or Underground Explosion. Boston, D. Reidel Publishing Co., 1981. p. 202. Copyright 1981 by John R. Murphy. Reproduced with the permission of John R. Murphy. All rights reserved.

The main strategies to defeat monitoring are to reduce the size of the seismic signal and to hide the explosion in other seismic noise. The most effective way to keep the signal small is to set off the explosion in a large cavity (i.e. an underground chamber) to muffle the conversion of its energy into seismic waves. This technique is termed "cavity decoupling." According to theory, decoupling can reduce the size of a seismic signal that an explosion produces by a factor of 10 to 50 for high-frequency seismic waves and by a factor of as much as 200 for low-frequency seismic waves.[11] This degree of decoupling for a 1-kt explosion would require a cavity at least 33 meters in diameter.[12] In the few decoupling experiments that have been conducted, however, the seismic signals have been reduced by only a factor of 70 at low frequencies and a factor of 10 at high frequencies.[13] Decoupling becomes more difficult for larger explosions. Larger explosions would require larger cavities; doubling the yield requires a cavity with double the volume. Larger cavities would be more difficult and expensive to construct, would be at more risk of collapsing following the explosion, and the increased construction activities would be more difficult to hide from reconnaissance satellites. In addition, the size of the seismic signal from a decoupled explosion still increases with the size of the explosion, making seismic detection likely for larger decoupled explosions. The size of explosions for which decoupling is credible, because of the size of the cavity needed, is probably limited in practical terms. Most seismologists put the upper limit for cavity decoupling at 10 kt, but some believe it may be as high as 30 kt.[14]

Decoupling could also be used to make explosions appear smaller than they actually are. For example, a 5-kt explosion could be made to appear in compliance with a 1-kt threshold by firing it in a cavity somewhat smaller than that which would be required for full decoupling. One way for a threshold treaty to avoid this problem might be to require that all tests be conducted at locations where cavity construction would be nearly impossible, such as in certain types of geology or at great depths. Another strategy would be to require on-site inspections of all tests. Short of this, seismic methods might help identify decoupling. Because decoupling is believed to be more effective at reducing low-frequency than high-frequency seismic waves, the seismic waves from decoupled explosions might be expected to have a higher fraction of high frequency waves.

A less effective way of reducing the seismic signal from an explosion is to set the explosion off in a weak and dry geologic deposit, such as dry alluvium, that would "muffle" the explosion. This can reduce the size of a seismic signal by about a factor of ten.[15] This method is not a major concern for hiding nuclear explosions in the Soviet Union for two reasons: (1) muffling in dry alluvium is much less potent than cavity decoupling for reducing the size of the seismic signal from an explosion. A seismic system that is designed to detect decoupled explosions of a given yield will easily detect muffled explosions of the same yield. (2) The Soviet Union is not believed to have sufficiently deep alluvium deposits to make muffling in alluvium attractive. The method would be more useful to confuse the estimation of yields than to hide explosions. To minimize the possibility of muffling, a treaty could allow testing only at sites that do not have the types of geologic deposits that are necessary for muffling. Such a requirement could restrict the United States more than the Soviet Union--the U.S. Nevada Test Site has extensive deposits of dry alluvium appropriate for muffling explosions.

Setting off an explosion at a time when there is high background noise, such as an earthquake, would also make detection more challenging, yet as illustrated in figure 5, monitoring at high frequencies has the potential to detect even small explosions set off in large earthquakes.[16] While currently deployed instruments do not provide this capability for the entire Soviet Union, it appears to be technically feasible.

Small decoupled nuclear explosions might be disguised as chemical explosions used in mining or construction.[17] Clear seismic discriminants between small nuclear and large chemical explosions have not yet been developed. One scenario for cheating would be to set off two chemical explosions at the same time, followed shortly by for example, a 5-kt decoupled nuclear explosion. The result would be two seismic signals, evidence that two chemical explosions had taken place, and possibly no reason to suspect that a nuclear test had occurred. While this approach might work, it would be difficult to run an extended test program in this manner, and suspicions would be aroused if the seismic signals from the chemical or nuclear explosions were seen as peculiar.

These evasive techniques, individually or in combination, increase the difficulty of monitoring testing of explosions with yields below ten kilotons. They increase the sensitivity required of a seismic monitoring system, increase the care with which treaties must be drafted, and increase the importance of nonseismic monitoring methods.

### *Nonseismic Monitoring Methods*

*Monitoring Underground Explosions.* Several nonseismic approaches may help detect testing-related activities or confirm the source of suspicious seismic signals.[18] Photoreconnaissance satellites can detect several types of activities or phenomena associated with testing. These include drilling and construction activities related to setting off underground nuclear devices, and surface effects, such as subsidence craters, that may result from tests. Photoreconnaissance satellites have sufficiently high resolution[19] to detect these activities if they are not hidden and if one knows where to look. Preparations for clandestine testing might be disguised as, for example, petroleum or mining activities. If a suspicious seismic signal is detected, satellites could examine the area for unusual activities and, if necessary, identify a specific location for an on-site inspection.[20] An on-site inspection might detect any radiation that reached the surface or evidence that suspicious activities had taken place. On-site inspections, however, might not confirm that a clandestine nuclear explosion had taken place. Surface evidence could be removed, and, if there were evidence, the violator could refuse to permit an on-site inspection to take place. Since a refusal could be seen as incriminating, on-site inspections may serve as a deterrent to violations.

The United States has detected radioactivity from several Soviet under ground tests that have vented to the atmosphere from airborne sampling. This is not a reliable method of detecting clandestine tests because they would undoubtedly be buried deeply to avoid venting radioactive gases to the atmosphere. Nevertheless, the possibility of error and detection adds to the risk a would-be violator faces.[21]

Other intelligence methods, such as electronic surveillance, defectors, or espionage, provide additional means of discovering test violations and would undermine a potential violator's confidence that it could test clandestinely. While no method is totally effective, each contributes to the likelihood of detecting clandestine testing. If the United States detects Soviet violations through these means, however, it may choose not to go public to avoid compromising the source of the information.

Nonseismic means to assure compliance with a threshold treaty include such cooperative measures as advance notification of tests and provisions for on-site observation and yield measurements. Onsite yield measurements are potentially more accurate than seismic methods but require the presence of one side's technical personnel at the other side's test sites.

An important nonseismic method of estimating the yield of explosions is CORRTEX ("Continuous Reflectometry for Radius Versus Time Experiment"). The Reagan Administration's policy was to support ratification of the TTBT if the Soviet Union agreed to an effective verification regime that included the use of CORRTEX.[22] CORRTEX consists of an electric coaxial cable placed in a hole close to the shaft containing the explosive (46 feet away is appropriate for 150 kt tests)[23] and associated recording devices (see figure 7). When the nuclear device is detonated, the shock wave from the explosion crushes the cable. The rate at which the cable is crushed, measured by the time that it takes electrical pulses sent down the cable to return, indicates the speed of the shock wave from the explosion, which in turn is used to calculate the yield of the explosion. The Peaceful Nuclear Explosions Treaty of 1976 provides a precedent for on-site observations of nuclear tests. It includes provisions for using a device similar to CORRTEX to measure

**Figure 7**

**CORRTEX**

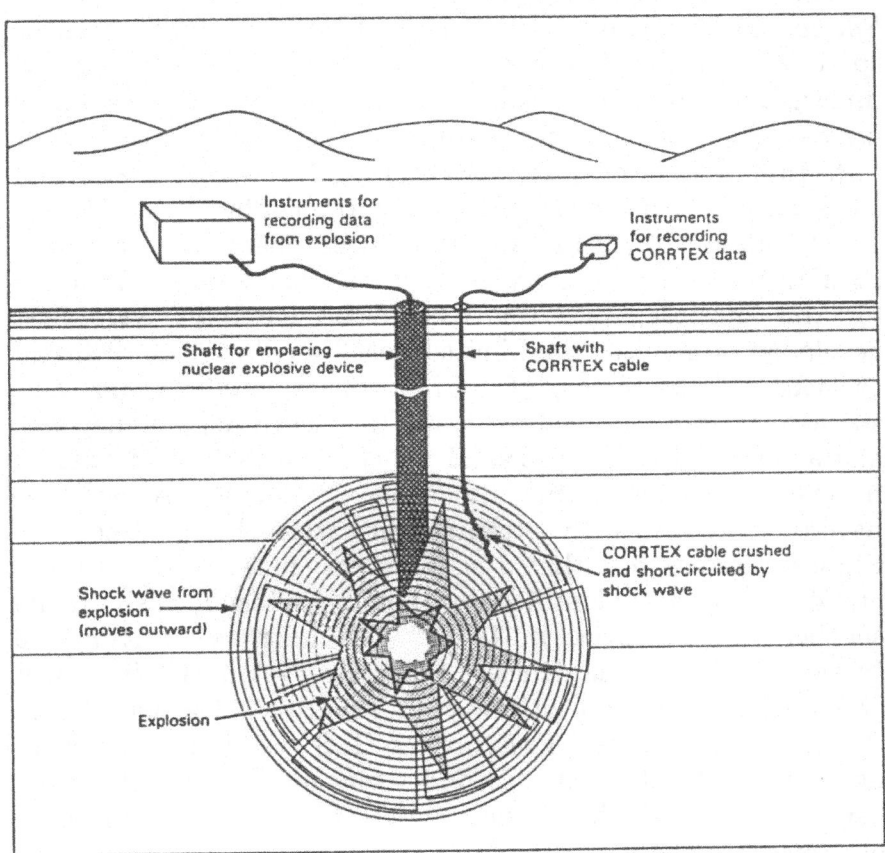

Distance between nuclear explosive shaft and CORRTEX shaft: 46 feet for a 150-kt explosive.

Depth of burial of nuclear explosive device: approximately 2,100 feet for a 150-kt explosive.

Source: Congressional Research Service. Based on diagrams by Los Alamos National Laboratory and the U.S. Department of State.

the yields of groups of explosions whose combined yield would exceed 150 kilotons.

***Monitoring Testing in the Oceans, Atmosphere, and Space.*** The possibility of testing in the oceans, atmosphere, or space is not of great concern as long as relatively unrestricted underground testing is allowed. If underground testing is increasingly restricted, there will be a greater potential benefit to testing in other environments and greater attention will need to be placed on monitoring these environments.

Detection of tests in the atmosphere is relatively straightforward. The United States has had a system in space for this purpose since the U.S. VELA satellites were first developed to monitor the Limited Test Ban Treaty of 1963 and the Outer Space Treaty of 1967.[24] These satellites and their successors were designed to detect the neutrons, gamma rays, and X-rays of tests in space and the characteristic light flash of nuclear explosions in the atmosphere.[25] The VELA satellites have been credited with detecting every atmospheric French and Chinese nuclear test that a satellite was in a position to see, and detected an unidentified, but possibly nuclear, flash off the southern tip of Africa in 1979.[26] An improved system, the Integrated Operational Nuclear Detection System (IONDS), will be launched over the next few years aboard the Navstar Global Positioning Satellites. The Navstar GPS system is planned to use at least 18 satellites, placed in orbits so that all parts of the earth's surface will be in the view of at least four satellites at all times.[27]

The Department of Energy asserts that the Soviet Union could test nuclear explosives in space behind a planet or the sun, or beyond the range of U.S. sensors, and send the results back to earth.[28] DOE estimates that such tests would cost on the order of $100 million, or about five times what most current nuclear tests cost. Others are skeptical about such a scenario. The Soviet Union sends relatively few (an average of less than three per year) space missions beyond earth orbit and the United States monitors all Soviet space launches.[29] In recent years, the Soviet Union has invited and obtained inter-national participation in its civilian space activities (the only space missions to go outside the Earth's orbit).[30]   Thus, a highly secretive Soviet mission into deep space would attract suspicion.   In addition, to obtain information from such a clandestine test, the Soviet Union would need to transmit information back to earth without it being detected by the United States.  This might not be done easily, particularly in view of the fact that the Soviet VEGA mission to Halley's comet required European and American radio telescopes to help track the spacecraft.[31]  Finally, if Soviet testing in space were a real concern, provisions might be written into a treaty to allow inspection of all spacecraft to be launched into regions of space where there could be a risk of undetected testing.

The oceans are generally a less attractive environment for clandestine testing than the continents.  Underwater tests can be monitored to levels far below the threshold for underground testing[32] using acoustic sensors, such as the Sound Surveillance System--a network of underwater hydrophones--used to track submarines.[33]  The sensitivity of such a system is determined by the number and sensitivity of the sensors.  Since there is no limit to the number of sensors permitted in the seas, the sensitivity of such a system is determined largely by economic considerations.  Identification of the country responsible for an underwater nuclear test, however, could be difficult to prove.

## Verifying Compliance With Specific Test Ban Treaties

### Verifying Compliance with the Threshold Test Ban Treaty and the Peaceful Nuclear Explosions Treaty

Although this report focuses on more restrictive test ban treaties than the TTBT and PNET, the debate over verification of these treaties is a barrier to considering more restrictive treaties.  The main issues concerning the verification of the these treaties are:  (1) Has the Soviet Union complied with the 150-kt limit of the TTBT?  (2) Do the verification provisions in the treaties provide for acceptable verification? (3) If verification needs to be improved before the treaties are ratified, what verification provisions should be added to the treaties? Each issue is the subject of controversy both within the Administration and the governmental seismic community, and between these groups and test ban supporters.

*The Issue of Soviet Compliance.* The Reagan Administration's position was that "Soviet nuclear testing activities for a number of tests constitute a likely violation of legal obligations under the TTBT of 1974."[34] Others believe that the evidence for these assertions is very weak, and that if a few Soviet tests have exceeded 150 kt, it has been within the range that the United States and Soviet Union agreed would be considered a serious matter but not violations of the treaty. The Soviet Union has asserted that all of its tests have had yields of less than 150 kt.

The center of the dispute has been the precise formula used to calculate the yields of Soviet nuclear explosions. As described previously, due to differences in the earth underneath the Nevada Test Site and test sites in the Soviet Union, an explosion at a Soviet test site produces a larger seismic signal than an equivalent explosion at the Nevada Test Site. At issue has been how large a correction factor, or "bias," to use when estimating the yield of Soviet explosions based on the experience with Nevada explosions. When the treaty first took effect, relatively little research had been conducted on how large this bias should be. Over time, research has supported a larger correction factor than that first used, and in 1985 several expert review panels reportedly recommended changing to a larger correction factor, which would reduce the yield estimates for Soviet tests.[35] While acceptance of this revised correction factor has reduced the range of technical debate, some seismologists, especially those who support further limitations on nuclear testing, advocate a somewhat larger correction that would further decrease the estimates of the yields of Soviet tests.[36]

With the revised correction factor, a few Soviet explosions are still estimated to be greater than 150 kt. This may be accounted for, however, by the uncertainty in yield estimates. Due to scatter in the data, some tests that are slightly under the threshold can be measured as being slightly above the threshold and vice versa. Some Soviet explosions that appear to be greater than 150 kilotons using one seismic method appear to be less than 150 kt when recalculated using a different seismic method.[37]

The position of Lawrence Livermore National Laboratory on the issue of Soviet compliance is:

> The Soviets appear to be observing some yield limit. Livermore's best estimate of this yield limit, based on a probabilistic assessment, is that it is consistent with TTBT compliance. However, because of the statistical uncertainty of teleseismic yield estimates and the uncertainty in extrapolating U.S. experience to Soviet test sites, we cannot rule out the possibility that a few Soviet tests may have exceeded the limit.[38]

Most seismologists familiar with the issue seem to agree with this view.[39]

A further consideration in the question of Soviet compliance or noncompliance with the 150-kt threshold is what is called the "whoops" agreement associated with the TTBT. When the TTBT was negotiated, the two sides were concerned that tests might accidentally produce yields greater than anticipated and thus exceed the 150-kt threshold. The United States and Soviet Union agreed that "one or two slight unintended breaches per year would not be considered a violation of the treaty," but would be a cause for concern that could be the subject for consultation on request.[40] No one asserts that the Soviets have breached the threshold on the average of more than twice a year. Whether breaches, if any, have been slight and unintended is more open to debate. Administration statements of likely noncompliance do not address this "whoops" agreement. Most seismologists believe it is quite possible that one or more Soviet tests have exceeded the 150-kt limit. Some weapon experts, however, would say that such violations have not been militarily significant.[41]

***The Adequacy of the Monitoring.*** The Reagan Administration's position was that "the TTBT and PNET are not effectively verifiable in their present form."[42] At issue is whether the uncertainty in yield estimates under the verification provisions in the treaty are acceptable.

The uncertainty in yield estimates when the TTBT and PNET were signed was commonly stated to be "plus or minus a factor of two,"[43] which means that if there were a large number of 150-kt tests, 95 percent of the yield estimates for those tests would be between 75 (i.e. 150/2) and 300 (i.e. 150 x 2) kilotons. Most of the yield estimates would fall towards the center of that range, and 5 percent would fall outside of the range. This is illustrated in figure 8a. The probability that the yield estimate for a 150-kt explosion will fall in any yield range is simply the percent of area under the curve for that yield range. The figure illustrates that, with a factor of two uncertainty, there is a 2.5 percent chance of a 150-kt explosion being measured as greater than 300 kt.

Although the Reagan Administration stated that the uncertainty from seismic measurement is still a factor of two,[44] many technical experts believe the uncertainty is somewhat less. Many seismologists cite a factor of 1.6 or 1.5.[45] The Administration viewed the uncertainties in estimating yields as unacceptable for verifying compliance with the TTBT, but has stated that uncertainty of a factor of 1.3, such as can be obtained using CORRTEX for explosions over 50 kt, is acceptable.[46] The difference between these factors is shown in figure 8b. As can be seen in the figure, as the uncertainty factor becomes smaller, the yield estimates for 150-kt explosions are likely to be much closer to 150 kt, and explosions whose yields are estimated to be greater than 150 kt can be determined with much higher confidence to in fact be above 150 kt.

What level of uncertainty is acceptable? The acceptable level of uncertainty depends on several factors, including the perceived military effects of violations and the negotiating posture with respect to future treaties. There is agreement that less uncertainty is better. With less uncertainty, fewer Soviet explosions of less than 150 kt are likely to be measured at levels that appear to be treaty violations, and the United States would have more confidence that explosions measured to be

## Figure 8

### Yield Estimates and Uncertainty Factors
### for 150-Kt Explosions

### Figure 8a

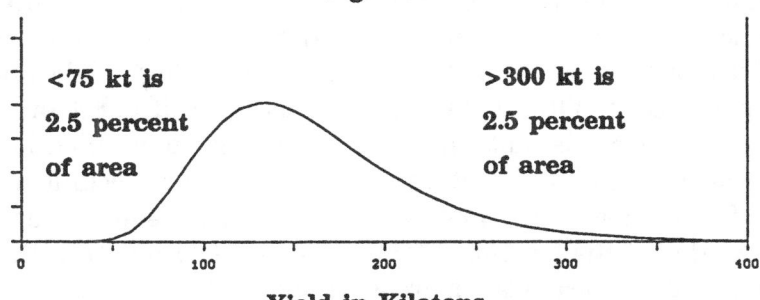

**Yield in Kilotons**

This curve indicates the distribution of yield estimates for 150-kt explosions when the uncertainty is a factor of two. The relative area under the curve for any yield range indicates the relative likelihood of the yield estimate for a 150-kt explosion falling in that range. For example, as indicated, there is a 2.5 percent chance that a 150-kt explosion will be measured to be greater than 300 kt or less than 75 kt.

### Figure 8b

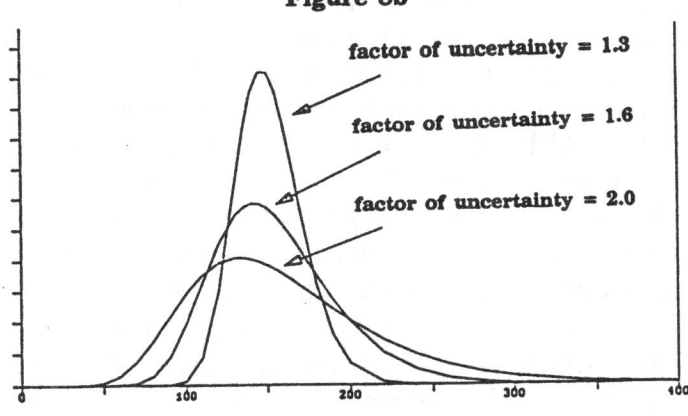

**Yield in Kilotons**

These curves show the difference between uncertainty factors of 2, 1.6, and 1.3. With a factor of 1.6 uncertainty, 150-kt explosions can be expected to be measured below 240 kt 97.5 percent of the time; with a factor of 1.3 uncertainty, 150-kt explosions can be expected to be measured below 195 kt 97.5 percent of the time.

Source:  Congressional Research Service.

greater than 150 kilotons were in fact treaty violations. At issue is whether better verification is necessary.

*How Should Verification Be Improved?* The Reagan Administration's position was that "we require . . . effective verification through direct, on-site hydrodynamic yield (CORRTEX) measurement of all appropriate high-yield nuclear detonations."[47] Others believe that using CORRTEX on every large explosion is not the best approach to verification for several reasons. It is intrusive, requiring U.S. technicians on the Soviet test site, and vice versa, and requires detailed knowledge of the location of the explosive device within the test shaft.[48] There are also increasing difficulties in using it for explosions with yields progressively less than 50 kt, perhaps creating a new barrier to verifying compliance with a lower yield threshold test ban.

Some people instead suggest that with improvements, seismic methods could result in uncertainties equivalent to those of CORRTEX. For example, unclassified sources indicate that seismic estimates of yields at the Nevada test site have an uncertainty of a factor of 1.3.[49] The advantages of using seismic methods are that they would be less intrusive and that they would work for much smaller explosions. Thus, TTBT verification procedures could be carried over to lower yield thresholds. Although it is not certain that seismic methods could achieve an uncertainty as low as a factor of 1.3 for Soviet tests, the current uncertainty would be reduced if the United States could calibrate its seismic methods for the Soviet test site and obtain better information on the test site geology. The seismic methods could be calibrated in one of several ways: (1) by using CORRTEX on several tests; (2) by putting seismographs on the Soviet test site; (3) or by setting off nuclear or chemical explosions of known yield on the Soviet test site. Better information on Soviet geology could be obtained by surface surveys and by rock samples from drill holes. Some of this information is being obtained from the activities in the Soviet Union of the Natural Resources Defense Council (NRDC). NRDC scientists have set up seismic monitoring stations in the vicinity of the main Soviet test site, and have set off chemical explosions in the Soviet Union to improve knowledge of the movement of seismic waves in the Soviet Union.[50]

Ratification of the TTBT would result in exchanges of some of this information, but the treaty does not contain provisions for the information to be independently verified. The information could be checked for consistency with information already available, but such checks are not likely to satisfy skeptics. If changes were made to the TTBT verification protocols to provide this information in a verified form, there could be greater confidence in verification.


### Verifying Compliance with a Comprehensive Test Ban Treaty

A CTBT has not been completed and thus verification procedures are not set. Therefore, the verification issue is whether a CTBT is potentially verifiable, and, if so, whether verification provisions could be acceptable to both sides. The Administration's position was that a complete test ban is not effectively verifiable.[51] There is general agreement that some small explosions would escape seismic detection by any practical monitoring system, and thus there will always be some possibility of clandestine tests under a CTBT. If possible Soviet violations that could go undetected are seen as having little military significance, however, and if the benefits of a CTBT are perceived to be great, a CTBT might still be seen as acceptably verifiable. The issues are: (1) What is the range of possible cheating on a CTBT in various hypothetical verification regimes? and (2) How militarily significant are these risks of cheating?

Any seismic monitoring system has a yield threshold below which it cannot detect or identify explosions. This limit depends largely on the design of the seismic network. With a very large number of seismic monitoring stations one can detect very small explosions. But because there are many small seismic events--such as blasting for mining and construction--the more sensitive the system, the more suspicious events will need to be identified. Thus, there is a practical limit to monitoring capabilities, due to the number of seismic stations within the Soviet Union that can be negotiated and the number of insignificant events that must be identified at the detection threshold.

Many seismologists agree that practical seismic monitoring networks that include stations within the Soviet Union could monitor nuclear explosions down to a range of one to a few kilotons even in the face of sophisticated attempts to cheat.[52] There is some dispute, however, about the specifications of the monitoring system required and what its exact capabilities would be. Some say that 25 single-seismograph stations within the Soviet Union complemented by 15 stations outside the Soviet Union could detect and identify 1-kt decoupled explosions with high confidence.[53] Others say that a similar number of more complex seismic arrays, each consisting of several seismographs, would be needed, and might still be unable to identify confidently all explosions of several kilotons.[54] There is agreement that further studies are needed to better understand the performance of these seismic monitoring networks, and that operations of these networks would require a substantial increase in the scope of the U.S. monitoring operations.

A good capability of monitoring 1-kt explosions, if achieved, could inhibit clandestine testing at some level less than 1 kt. If a network has the capability to detect and identify decoupled 1-kt explosions with high confidence anywhere in the Soviet Union, the network would be able to detect and identify decoupled explosions at some level less than 1 kt with somewhat less confidence. Explosions set off in hard rock (not decoupled) could be detected with high confidence to a considerably lower level. Other sources of intelligence, including reconnaissance satellites, electronic surveillance, and human intelligence, would also make extensive clandestine testing at lower levels risky.

On the other hand, many small seismic events, largely chemical explosions and small earthquakes but possibly decoupled nuclear explosions of a few kilotons, would be detected that could not be positively identified. It is likely that some of these could not be eliminated as possible nuclear tests and thus might create concern that the Soviet Union was conducting small nuclear tests.

Whether these monitoring capabilities are acceptable for monitoring a "comprehensive" test ban would depend heavily on the perceived military significance of low yield explosions. Small nuclear tests can be useful for effects testing, for conducting research into new weapons, and for developing new weapons of low yield, but it is not clear if clandestine testing at less than 1 kt would lead to militarily significant new weapons. Small nuclear tests might be significant as preparations to break out of a treaty. They could help a country maintain weapons expertise and support research and development activities that could lead to new weapons more quickly after breaking out of a treaty.

The precise definition of a CTBT would influence the extent to which there would be potential asymmetries. For example, if a CTBT were to allow testing inside a reusable laboratory, both sides could continue research and testing at yields up to as much as a few tenths of a kiloton. This could reduce the military advantage, and thus the incentive, of conducting small clandestine decoupled explosions beyond the bounds of the treaty. Such a broad definition of a CTBT, however, would make it in effect a low-yield threshold treaty.

What kind of verification regimes can be negotiated? In recent treaty negotiations the Soviet Union has agreed to on-site monitoring and inspections. Provisions of the PNET call for the presence of observers and measurements using a CORRTEX-type device to ensure that no test in a group of peaceful nuclear explosions exceeds 150 kt.[55] During the 1978-1980 CTBT negotiations, the Soviet Union agreed in principle to 10 unmanned seismic observatories upon its soil.[56] The recently signed treaty eliminating intermediate range nuclear forces (INF Treaty) has provisions for extensive on-site monitoring. In 1986, the Soviet Union accepted three seismographs installed and operated near its main test site by the Natural Resources Defense Council, a private U.S. organization, to monitor the Soviet testing moratorium. The Soviet Union's stated policy is that verification is not a problem and that it is open to international verification measures.[57] In December of 1987, the United States and the Soviet Union agreed to conduct a joint verification experiment consisting of one nuclear explosion at each side's test site to calibrate seismic methods of verification and to demonstrate the use of CORRTEX. It remains to be seen whether a large number of in-country monitoring stations and meaningful rules for on-site inspections could in fact be negotiated.

### Verifying Compliance with a Low-Yield Threshold Treaty

Because a low-yield threshold treaty has not received as much attention as the CTBT or the TTBT, verification of these treaties has not been as widely discussed. As discussed in previous chapters, two thresholds of possible interest may be: (1) a threshold of 20 kt; and (2) a threshold of 1 kt that would be a substitute for a comprehensive test ban. For each threshold, the verification issues are: (1) What seismic monitoring system is needed to detect and identify significant explosions outside of the designated test sites? (2) What verification provisions are necessary to determine, with sufficient certainty, if the yields of explosions at designated test sites are within the threshold? (3) Are these verification provisions negotiable?

Verifying compliance with a 20-kt threshold would be similar to verifying compliance with the current 150-kt threshold. Detection and identification of significant explosions (explosions with yields close to or greater than the threshold) outside of designated test sites would not be a problem at the 20-kt level anywhere in the Soviet Union with the current monitoring system. Hiding explosions as large as the threshold with cavity decoupling would be impractical and the detection threshold for nondecoupled explosions is well below 20 kt. Estimating yields would also differ little from the TTBT.[58] Everything else being equal, the uncertainty in yield estimates for tests at 20 kt would be somewhat greater than for tests at 150 kt.[59] If verification provisions in a treaty allowed for calibration of the seismic methods and the test sites, as well as a few in-country seismic monitoring stations near the test sites, it could be possible to estimate the yields of 20-kt explosions with about the same uncertainty as could be possible by seismic means for the current 150-kt threshold.

A 1-kt LYTT would be considerably more difficult to monitor and substantial technical issues are unresolved. Detection and identification of clandestine explosions would require an in-country monitoring network similar to that envisioned for a comprehensive test ban. Estimating yields around 1 kt would also be more difficult than at 20 kt. It would be desirable to have all the verification provisions described above for a 20-kt threshold, including several seismic monitoring stations near the test site to ensure that strong seismic signals could be picked up for small explosions. An additional difficulty is that it is possible to use partial decoupling or muffling to confuse the estimation of the yield of small explosions, and thus make an explosion of several kilotons appear to be less than one kiloton. If cheating were possible in the 1- to 5-kt range, it could be of military significance. Therefore, a treaty would need to ban decoupling or muffling at the test site and have provisions to enforce this, such as a requirement for explosions to be at great depth or in certain types of geology, or on-site inspections. These problems have not been widely addressed. It is not clear what level of uncertainty would be acceptable or negotiable in estimating the yields of explosions around 1 kt.

### Verifying Compliance with a Quota Treaty

As with a low threshold test ban, the difficulties of verifying compliance with a Quota Treaty have not been widely publicly discussed. A quota refers to a limit on the numbers of tests, but would likely be combined with an upper threshold and perhaps a lower threshold. The upper threshold could be the current 150-kt threshold or some lower (or conceivably higher) threshold. A lower threshold might be set at some low verifiable level to prevent the possibility of asymmetrical clandestine testing at levels below the detection threshold. To verify compliance with a Quota Treaty, one needs to verify observance of upper and lower thresholds (if any) as well as count the number of tests. Verifying compliance with these thresholds would be the same as described earlier.

The potential difficulty in counting tests would be to determine if multiple devices were tested simultaneously in the same or a neighboring hole. For example, two 75-kt devices exploded simultaneously could produce a seismic signal similar to one 150-kt shot. There are several possible approaches to this problem. One would be to allow testing of multiple explosions within a ceiling on total yield for a test and let the tester choose between one large device and several smaller devices in a test. If a quota were combined with a lower threshold, these trade-offs would be more difficult. For example, there may be less military advantage to testing two 5-kt devices in a permitted 10- kt test than to testing ten 15-kt devices in a permitted 150-kt test. It is also possible that some type of on-site inspections could be developed to prevent testing several nuclear devices in one test.

## Notes

1. Earle, Ralph, III. Verification Issues from the Point of View of the Negotiator. Chapter 3 in Tsipis, Kosta, David W. Hafemeister, and Penny Janeway, eds. Arms Control Verification: The Technologies that Make It Possible. Washington, Pergamon-Brassey's, 1986. p. 16.

2. Perle, Richard, Kenneth Adelman, and Robert Barker. Response to questions submitted by Senator Carl Levin. In U.S. Congress. Senate. Committee on Armed Services. Nuclear Testing Issues. Hearings. 99th Cong., 2nd sess., April 29 and 30, 1986. Washington, Govt. Print. Off., 1987. p. 45.

3. See also U.S. Office of Technology Assessment. Seismic Verification of Nuclear Testing Treaties. Washington, U.S. Govt. Print. Off., May 1988. 139 pp. This report, published while this chapter was undergoing reviews, comprehensively covers seismic verification issues.

4. For seismic waves, "high frequency" waves are those with frequencies of more than 3-5 hertz (cycles per second).

5. For a description of some of these discriminants, see the following articles in Kerr, Ann U., ed. The VELA Program: A Twenty-Five Year Review of Basic Research. Arlington, VA. Defense Advanced Research Project Agency, 1985: Marshall, P.D., and A. Douglas. Earthquake or Explosion: Teleseismic Monitoring--Where Are We Now? p. 633-657; and Blandford,

Robert. Regional Detection, Location, Discrimination and Yield Determination. p. 787-816.

6. This would be much like using a "graphic equalizer" found on many stereo systems to eliminate low frequency and accentuate high frequency sounds.

7. Although this method is one way of distinguishing explosions from earthquakes, it is not the most commonly used method. Most explosions are distinguished from earthquakes on the basis of their location and the relative magnitudes of their surface and body waves. In this case, the fact that the earthquake was farther away from the monitoring station than the explosion also helped to distinguish the explosion from the earthquake because the earthquake's high frequency waves were attenuated by the distance.

8. Magnitude is basically a measure of the amplitude of the seismic waves from an event, corrected for various factors, such as distance, so that all seismographs that detect an event will assign it approximately the same magnitude. It is measured on a logarithmic scale, so that an increase of one in magnitude represents an increase of a factor of ten in ground motion and about a factor of 30 in energy released.

9. Surface waves are not distorted in the same way. For details, see CRS Report No. 86-155 SPR, Monitoring Nuclear Test Bans.

10. Richards, Paul G. Tutorial on Estimation of Nuclear Explosion Yield Using Seismic Methods. Unpublished paper, 1987. 33 p.

11. Evernden, J.F., C.B. Archambeau, and E. Cranswick. An Evaluation of Seismic Decoupling and Nuclear Test Monitoring Using High Frequency Seismic Data. Reviews of Geophysics, v. 24, no. 2, May 1986: 143-215.

12. Ibid. Others suggest larger cavities are required for decoupling. For example, W.J. Hannon says cavities of 40,000 to 100,000 cubic meters are needed to decouple a 1-kt blast, which corresponds to cavities with diameters of 42 to 58 meters.

13. Evernden, J.F., C.B. Archambeau, and E. Cranswick. An Evaluation of Seismic Decoupling and Nuclear Test Monitoring Using High Frequency Seismic Data. Reviews of Geophysics, v. 24, no. 2, May 1986: 143-215.

14. Ralph Alewine of the Defense Advanced Research Projects Agency suggested (telephone interview, June 21, 1986) that it might be possible to decouple a 30-kiloton explosion, but other seismologists put the upper limit at 10 kilotons (Willard Hannon, Lawrence Livermore National Laboratory, telephone interview, November 1986; Paul Richards, Lamont-Doherty Geological Observatory of Columbia University, personal interview, December 1986).

15. Hannon, W.J. Seismic Verification of a Comprehensive Test Ban. Science, v. 227, Jan. 18, 1985: 251-257.

16. For a more complete discussion of discriminating between explosions and earthquakes, see Evernden, J.F., C.B. Archambeau, and E. Cranswick. An Evaluation of Seismic Decoupling and Nuclear Test Monitoring Using High Frequency Seismic Data. Reviews of Geophysics, v. 24, no. 2, May 1986: 143-215.

17. For a discussion of the problem of discriminating between chemical and nuclear explosions, see Richards, Paul G. Consideration of Chemical

Explosions. Presentation to the Office of Technology Assessment, April 28, 1987. 7 p.

18. A more complete description of many of these techniques can be found in Zorpette, Glenn. Monitoring the Tests. IEEE Spectrum, July 1986: 57-66.

19. By one report, photoreconnaissance satellites have a resolution of about 10 centimeters (4 inches). Hafemeister, David, Joseph J. Romm, and Kosta Tsipis. The Verification of Compliance with Arms-Control Agreements. Scientific American, v. 252, March 1985: 39-45.

20. For a discussion of on-site inspection, see Shearer, Richard L., Jr. On-site Inspection for Arms Control. Washington, National Defense University Press, 1984. 69 p.

21. In the U.S. experience, the risk of leakage is greater at low yields because the explosion is too weak to melt a sufficiently thick layer of rock and earth to seal off the underground cavity. Paul Brown, Lawrence Livermore National Laboratory. Personal communication with David Cheney.

22. Verifying Nuclear Testing Limitations: Possible U.S.-Soviet Cooperation. U.S. Department of State, Bureau of Public Affairs, Special Report No. 152. August 14, 1986. p. 7.

23. Ibid. p. 5.

24. U.S. Congress. House. Committee on Science and Technology, Subcommittee on Space Science and Applications. United States Civilian Space Programs, 1958-1978. 97th Cong., 1st sess. Washington, Govt. Print. Off., 1981. p. 899.

25. Smith, R. Jeffrey. High-tech Vigilance. Science 85, December 1985: 31-32.

26. Ad Hoc Panel Report on the September 22 Event. Washington, Executive Office of the President, Office of Science and Technology Policy. 1980. 20 p.

27. Ibid., p. 64

28. U.S. Department of Energy. Nuclear Weapons Testing. Policy Paper 5, January 1987. Also comments on report drafts by Lawrence Livermore National Laboratory, July 1987.

29. For an overview of the Soviet space program, see U.S. Library of Congress. Congressional Research Service. Space Activities of the United States, Soviet Union and Other Launching Countries: 1957-1986. Report No. 87-229 SPR by Marcia S. Smith. February 27, 1987.

30. Waldrop, M. Mitchell. Soviet Space Science Opens to the West. Science, June 12, 1987: 1427-1431.

31. Waldrop, Soviet Space Science Opens to the West, p. 1430.

32. National Academy of Sciences. Committee on International Security and Arms Control. Nuclear Arms Control: Background and Issues. Washington, National Academy Press, 1985. p. 216.

33. Adam, John A. Counting the Weapons. IEEE Spectrum, July 1986: 46-56.

34. U.S. Policy Regarding Limitations on Nuclear Testing. Washington, U.S. Department of State, Bureau of Public Affairs, Special Report No. 150. August 1986: 2.

35. According to a report in Science, the Department of Defense Technical Review Panel on Nuclear Test Ban Verification, which included both government and academic seismologists, reached a strong consensus in favor of revising the correction factor. A separate Air Force Technical Applications Center panel reached a similar conclusion. See Smith, R. Jeffrey. Dispute Over Soviet Testing Heats Up. Science, v. 228, May 31, 1985: 1072. Also: Joyce, Christopher. New Method Lowers Yield of Soviet Tests. New Scientist, April 17, 1986: 15. Also: Gordon, Michael R. C.I.A. Changes Way That It Measures Soviet Atom Tests. New York Times, April 2, 1986: 1, 10.

36. Evernden, Jack F. Regional Bias in Magnitude versus Yield Measurements: Its Explanation and Modes of Evaluation. In: U.S. Congress. House. Committee on Foreign Affairs. Subcommittee on Arms Control, International Security and Science. Proposals to Ban Nuclear Testing. Hearings. 99th Cong., 1st sess., February 20; April 30; and May 8, 14, and 15, 1985. Washington, Govt. Print. Off., 1986. p. 194.

37. Sykes, Lynn R. and Ines L. Cifuentes. Yields of Soviet Underground Nuclear Explosions From Seismic Surface Waves: Compliance With the Threshold Test Ban Treaty. Proceedings of the National Academy of Science, v. 81, March 1984: 1922-1925.

38. Statement of Roger E. Batzel, Director of Lawrence Livermore National Laboratory, in U.S. Congress. Senate. Committee on Foreign Relations. Threshold Test Ban Treaty and Peaceful Nuclear Explosions Treaty. Hearings. 100th Cong., 1st sess., Jan. 13 and 15, 1987. Washington, Govt. Print. Off., 1987. p. 221.

39. Author's conversations with numerous seismologists, including current and former Department of Defense employees and contractors, employees of other U.S. Government agencies, and academic seismologists.

40. National Academy of Sciences, Committee on International Security and Arms Control. Nuclear Arms Control: Background and Issues. Washington, National Academy Press, 1985. p. 198.

41. Hannon, Willard J. April 29, 1985, letter to Congressman Dante B. Fascell. Hannon said that based on his conversations with weapons designers, he concluded that potential Soviet violations were not significant. In: U.S. Congress. House. Committee on Foreign Affairs. Subcommittee on Arms Control, International Security and Science. Proposals to Ban Nuclear Testing. Hearings. 99th Cong., 1st sess., February 26, April 30, and May 8, 14, and 15, 1985. Washington, Govt. Print. Off., 1986. p. 224-225.

42. Ronald Reagan, letter to the U.S. Senate. Congressional Record, Daily Edition, v. 133, January 13, 1987: S 685.

43. Giller, Edward B. Technical Issues Related to Test Ban Treaties. In Kerr, Ann V., ed. The VELA Program: A Twenty-Five Year Review of Basic Research. Arlington, VA, Defense Advanced Research Project Agency, 1985. p. 31.

44. See, for example, statement of Robert Barker in U.S. Congress. Senate. Committee on Foreign Relations. Threshold Test Ban Treaty and Peaceful Nuclear Explosions Treaty. Hearings, 100th Cong., 1st sess., Jan. 13 and 15, 1987. Washington, Govt. Print. Off., 1987. p. 89.

45. A factor of two uncertainty is based only on measurements from one type of seismic wave. When information from several types of seismic waves is combined, the uncertainty becomes less. For example, Paul Richards cited a factor of 1.5, in testimony before U.S. Congress. Senate. Committee on Foreign Relations. Threshold Test Ban Treaty and Peaceful Nuclear Explosions Treaty. Hearings, 100th Cong., 1st sess., Feb. 26, 1987. Washington, Govt. Print. Off., 1987. p. 85.

46. U.S. Department. of State. Bureau of Public Affairs. U.S. Policy Regarding Limitations on Nuclear Testing. Special Report No. 150. Washington, August 14, 1986. 6 p.

47. Ronald Reagan, letter to the U.S. Senate. Congressional Record, Daily Edition, v. 133, January 13, 1987: S 685.

48. Some limitations of CORRTEX are discussed in Lamb, Frederick K. Monitoring Yield of Underground Nuclear Tests. Reprinted in the Congressional Record, Daily Edition, v. 133, February 19, 1987: S 2357-S 2361.

49. Richards, Paul G. Information submitted for the record. U.S. Congress. Senate. Committee on Foreign Relations. Threshold Test Ban Treaty and Peaceful Nuclear Explosions Treaty, p. 358; see also CRS Report No. 86-155 SPR.

50. Nuclear Glasnost. The Amicus Journal, vol. 9, Fall 1987: 14-19.

51. U.S. Department of State. Bureau of Public Affairs. Verifying Nuclear Testing Limitations: Possible U.S.-Soviet Cooperation. Special Report No. 152. Washington, August 14, 1986. 7 p.

52. Sykes, Lynn R. and Jack F. Evernden. The Verification of a Comprehensive Nuclear Test Ban. Scientific American, v. 27, October 1982: 47-55; Hannon, W.J. Seismic Verification of a Comprehensive Test Ban. Science, v. 227, January 18, 1985: 251-257. Hannon believes now that explosions can be monitored to a lower level than when the paper was written. Hannon, W.J. Personal communication, November 1986; Evernden, J.F., C.B. Archambeau, and E. Cranswick. An Evaluation of Seismic Decoupling and Nuclear Test Monitoring Using High Frequency Seismic Data. Reviews of Geophysics, v. 24, May 1986: 143-215.

53. Ibid. Evernden, J.F., et al.

54. Hannon, W.J. Personal communication, November 1986.

55. Protocol to the Treaty Between the United States of America and the Union of Soviet Socialist Republics on Underground Nuclear Explosions for Peaceful Purposes. Appendix F in National Academy of Sciences. Committee on International Security and Arms Control. Nuclear Arms Control: Background and Issues. Washington, National Academy Press, 1985. p. 350-362.

56. National Academy of Sciences. Committee on International Security and Arms Control. Nuclear Arms Control: Background and Issues. Washington, National Academy Press, 1985. p. 200.

57. Gorbachev, Mikhail. Statement on Soviet television, August 18, 1986. Moscow. Novosti Press Agency Publishing House. 1986. 15 p.

58. Richards, Paul G., and Allan Lindh. Toward a New Test Ban Regime. Issues in Science and Technology, Spring 1987: 101-108.

59. Some seismic waves from smaller tests will be recorded at fewer seismic stations, reducing the amount of information obtained from smaller tests and increasing the uncertainty.

# 8

# Effects of More Restrictive Test Bans on Other Countries·

*Warren H. Donnelly, Charlotte P. Preece,*
*and Robert G. Sutter*

As nuclear weapons and weapons-development capabilities proliferate, any U.S.-Soviet test ban negotiations increasingly will have to take into account the concerns and behavior of other nations. Willingness of other nuclear weapons states and non-nuclear weapons nations with strong nuclear industrial bases to participate in a more restrictive test ban regime will be essential to the long-term success of global efforts to constrain the spread of nuclear weapons.

Nonnuclear weapons states with little potential or interest in developing nuclear weapons, in general, favor a comprehensive test ban as a step towards ending the nuclear arms race and as a powerful vehicle to dissuade those nations close to developing nuclear weapons from acquiring such arsenals.

The first section of this chapter examines the attitudes and policies of the major non-superpower nuclear weapons nations--France, Great Britain, and the People's Republic of

---

*Charlotte P. Preece and Robert G. Sutter prepared the section on "Attitudes of Third World Country Nuclear Powers and Nonnuclear Allies Toward a Test Ban." Warren H. Donnelly prepared the section on "Some Implication of More Restrictive Test Bans for the Non-Proliferation Regimes."

China--towards nuclear testing and various types of test bans. It also considers the views of several nonnuclear NATO powers toward a U.S.-Soviet comprehensive test ban. The second section draws out some of the implications of more restrictive test bans for nonnuclear weapons states--including those nations now considered to be at or near the threshold of producing nuclear weapons--and for U.S. efforts to extend the Non-Proliferation Treaty into the twenty-first century on terms favorable to U.S. security interests.

## Attitudes of Third Country Nuclear Powers and Nonnuclear Allies Toward a Test Ban

One condition that might affect a U.S.-Soviet decision to sign a CTBT or a more restrictive test ban agreement is the willingness of other nuclear weapons states to accept similar constraints. The United Kingdom, France, and the People's Republic of China have conducted 41, 141, and 32 tests respectively from the inception of their nuclear weapons programs through December 31, 1987. At present, all these countries assert that nuclear deterrence is the foundation for their security; that their nuclear arsenals contribute to global as well as national deterrence; that unless the superpowers first agree to deep cuts in strategic nuclear weapons, they must maintain and/or modernize their nuclear arsenals to retain minimal deterrent capabilities; and, that testing is vital to ensuring the reliability of their deterrents. In other words, these three nuclear weapons states view a test ban as the result of superpower agreement on significant arms reduction and not as a vehicle to create future arms reduction for themselves or others.

Despite general agreement opposing a CTBT at this time, the United Kingdom, France, and China conduct quite different testing programs, which probably contributes to their differing attitudes on nuclear testing restrictions. This chapter of the study briefly describes the nuclear testing policies of France, the United Kingdom, and China, and explores these governments' positions toward comprehensive and limited bans on nuclear testing. It concludes with an assessment of attitudes of America's non-nuclear allies toward a test ban.

## French Attitudes Toward a Test Ban

*French Nuclear Testing Policy.* To date, the government of France has conducted 141 nuclear tests since it exploded its first nuclear device in 1960.[1] France began its nuclear testing program with atmospheric bursts over the Algerian desert. Since 1966, France has conducted the remainder of its nuclear tests in the South Pacific atoll of Mururoa. France is not a signatory of any test ban treaties, and continued atmospheric testing atmospheric until September 1974--ten years after the United States, the Soviet Union and Great Britain stopped above-ground detonations. France is currently embarked on a vigorous strategic and tactical nuclear modernization program that will not be completed until the late 1990s. Since 1980, France has conducted 6 to 12 tests per year. Although public information on the size of French nuclear tests is not readily available, it is widely believed that French tests have in recent years been confined to well under the 150 kt limit set by the Threshold Test Ban Treaty, to which France is not a signatory. When asked if France plans to increase the number of tests in future years to meet modernization requirements, a French government official refused to comment.[2] France reportedly conducted 8 nuclear tests in 1987.[3]

Although the French government has not preannounced nuclear tests, it has on several occasions invited observers from neighboring Pacific nations to monitor tests in order to confirm their environmental safety. The New Zealand observatory at Rarotonga on the Cook Islands normally publicizes French tests, after which they are usually confirmed by the French government. Following the June 12, 1988, parliamentary elections, Foreign Minister Roland Dumas remarked to the UN third special session on disarmament that France will henceforth announce at the end of each year the number of nuclear tests it has conducted over the previous 12 months.[4]

*French Government Attitudes Toward Restrictions on Nuclear Testing.*[5] The official French government position on continuing French nuclear testing is that testing is an absolute essential as long as offensive nuclear weapons remain the foundation for deterrence. Government officials maintain that it is vital to continue to test to ensure the safety and reliability of existing nuclear systems as well as to develop new systems. Despite the French government's firm position on the need for testing, French officials are not opposed to limiting or banning nuclear testing as a part of a long-term arms control effort. Nonetheless, the government does not deem it appropriate to ban or limit testing until: (1) the superpowers make deep cuts in their strategic weapons; (2) the NATO/Warsaw Pact imbalance in conventional forces and chemical weapons is restored; and (3) the superpowers maintain adherence to the ABM Treaty, thus, in the French view, ruling out significant improvements in defensive systems. The French government does not accept the relationship that some arms controllers advocate of decreasing the number of tests as national nuclear arsenals are reduced. They argue that such limits on testing perpetuate the advantage of the superpowers in both numbers and quality of weapons and tests. Therefore, at this time the French would not agree to any type of limited or comprehensive ban on French nuclear tests.

Going beyond France's national interest in refusing to participate in a test ban, the French embassy official interviewed for this section stated that it would be unwise for the United States to sign a CTBT at this point. He said that if the United States ended nuclear testing it would cast doubts on U.S. commitments for extended deterrence to Western Europe. If the United States agreed to a quota test ban or a lower threshold ban, he maintained that it would have little effect on French testing policy other than to increase inter-national pressures on France to change its testing policy. He noted that such pressure would probably have little effect on French testing because of the high degree of domestic public consensus behind French security policy. On February 6, 1987, the United States announced that it would not sign the protocol to the South Pacific Nuclear Free Zone Treaty--a pact among 13

South Pacific island nations, including Australia and New Zealand. The Treaty nations asked the five major nuclear powers to sign protocols of association with the treaty that would forbid them from testing, using or threatening to use nuclear weapons in the area. The Soviet Union and the People's Republic of China signed these protocols in early 1987. Although the United States has supported nuclear-free-zones in Latin America, Antarctica and the ocean seabed, the Reagan Administration declined to sign the protocols "under current circumstances," citing strong pressure exerted by the French government and general concern over the implications of creating additional nuclear-free-zones in areas of special interest to the United States.[6]

### British Attitudes Toward a Test Ban

**British Nuclear Testing Policy.** The government of the United Kingdom has conducted about 41 nuclear tests since it exploded its first nuclear device in 1951.[7] The U.K. began its nuclear testing program with atmospheric explosions over the Australian desert. Since 1962, all British nuclear tests, singly or jointly with the U.S., have been conducted in the United States at the Nevada Test Site. Since the United States and the Soviet Union agreed in 1974 not to conduct any underground nuclear tests that exceed a yield of 150 kt, it can be assumed that the British tests conducted in the U.S. meet these criteria as well. As of January 1, 1988, the United States had conducted 20 joint tests with the U.K.[8]

Like France, Britain is also pursuing a vigorous strategic nuclear modernization program designed to replace its aging Polaris SLBM force by the mid-1990s with the Trident II (D-5 missile) system. Britain will construct the submarines and warheads for the system while the United States will sell Britain the D-5 missiles. Since 1980, the United States has conducted 1-to-3 tests per year for the British. D-5 testing is included in the price the United Kingdom will pay to the United States for developing the missile. It is not known whether British modernization requirements will demand more frequent testing in future years. Presumably, since the United States is performing its own development tests on the D-5, Britain could continue to undertake a relatively low number of tests over the next decade. The British government does not announce all nuclear tests.

***British Government Attitudes Toward Restrictions on Nuclear Testing.*** Prime Minister Margaret Thatcher has consistently made it clear to President Reagan that Britain has no intention of scrapping or curbing its independent nuclear force unless the Soviets sharply reduce their nuclear and conventional forces in Eastern Europe. Prime Minister Thatcher sought to temper the Reagan Administration's earlier enthusiasm for eliminating all strategic ballistic missiles by pointing out that such a proposal would leave Western Europe vulnerable to superior Warsaw Pact conventional forces.[9] At the same time, British Foreign Secretary Geoffrey Howe told Soviet Foreign Minister Eduard Shevardnaze that no country could accept the abolition of nuclear weapons within ten years as long as Soviet superiority in conventional and chemical weapons exists.[10] According to George Younger, British Secretary of State for Defence, Britain would be prepared to reexamine its nuclear capabilities provided the superpowers cut their strategic nuclear arsenals "very substantially" below the 50% reduction proposed in the START negotiations.[11]

In this context, the British--like the French--reject a CTBT, citing the necessity of testing for new weapons development, reliability and safety. At the last Non-Proliferation Treaty Review Conference in 1985, Richard Luce, British Minister of State at the Foreign Office, said: "We believe that a CTBT which allowed any militarily significant cheating by its signatories would not be in the interest of national stability or security."[12] Nonetheless, the United Kingdom was the third party to participate in talks on a CTBT that were suspended by the Reagan Administration in 1982. The British government supports the step-by-step approach to nuclear testing constraints advocate by President Reagan in 1986. Reagan's proposal, which led to U.S.-Soviet negotiations in September 1987, has as a first objective to develop verification arrangements that would allow for U.S. ratification of the TTBT and the PNET. The ultimate goal of U.S. policy is a CTBT.[13] Official British policy holds that a CTBT is desirable, but that the problem of verification remains an impediment to an agreement.[14]

The Thatcher government would probably not view a U.S.-Soviet CTBT in Britain's best interest for the following reasons: (1) It believes that testing is essential to ensure a credible deterrent; (2) It would mean that the British would have to forgo plans to deploy Trident or look elsewhere--perhaps to France--to conduct their nuclear tests; and, (3) It would undermine confidence in the reliability of America's nuclear guarantee to Western Europe. Concerning limited test bans, the Thatcher government reportedly would like to see the permissible nuclear threshold to be reduced gradually.[15]

Britain has no official position on a quota test ban. Given the aforementioned coincidence of U.S. and British positions on nuclear policy, the Thatcher government would probably deem it inappropriate to take a position on a quota test ban at least until the time when and if the U.S. Government decides to pursue such an objective. On the other hand, given its commitment to the necessity of modernizing its strategic forces, the Thatcher government would likely disapprove of any limits placed on U.S. testing that would jeopardize America's ability to continue joint U.S.-British testing.

## *Chinese Attitudes Toward a Test Ban*

*Chinese Nuclear Testing Policy.* The Chinese government is reported to have conducted 32 nuclear tests since it exploded its first nuclear device in the atmosphere in 1964.[16] For the past several years China has conducted, on the average, 1 test every two years. Due to China's deliberate pace of nuclear modernization, it is unlikely that China will markedly increase its number of nuclear tests over the next decade.

Most Chinese nuclear explosions have been conducted at a testing range in Northwestern China. China has performed both underground and atmospheric tests. In March 1986, the government announced that henceforth all nuclear tests would be underground. China has conducted high-yield and low-yield tests, in some cases exceeding 1 Mt.

Until 1980, China regularly announced its nuclear tests. The Chinese government has not announced a test since that time. China does not permit foreign inspections of its nuclear tests. While the Beijing government voices general support for "necessary" means to verify compliance with international arms control agreements, it is unclear if this would translate into specific Chinese willingness to allow foreign observers to make on-site inspection.

***Chinese Government Attitudes Toward Restrictions on Nuclear Testing.*** The Chinese government does not have a well developed position on issues related to the establishment of nuclear test ban agreements. Consultations with informed U.S. and Chinese observers reveal a current Chinese posture on these issues that is crafted generally to support important Chinese security interests and China's recent emphasis in support of international arms control.[17]

China's policy in this area is governed by several important objectives. First, China wants to maintain considerable freedom in developing its nuclear capabilities against likely adversaries. For over 20 years, Chinese leaders have given a high priority to Chinese nuclear weapons development in order to help deter outside aggression and intimidation, to secure a strategic retaliatory capability, and to demonstrate China's international importance. Since the early 1970s, China's deterrence and strategic retaliatory capabilities have focused on the Soviet threat. Beijing has deployed over 100 missiles, and perhaps some bombers, capable of hitting Soviet targets with nuclear warheads. It wants to remain reasonably sure that the Soviet Union will continue to be unable toneutralize this force with a first strike against China.

Additionally, China wants to exert influence on the U.S.-Soviet arms control negotiations. In the 1970s, Beijing encouraged Western resistance to Soviet overtures in arms control, partly in order to avoid a decline in U.S. military pressure on the Soviet Union that would allow Moscow greater freedom in dealing with its China problem. In the 1980s, as

Beijing became less concerned about possible U.S.-Soviet collaboration against China, it adopted a different approach. It endorsed a more moderate stance on U.S.-Soviet arms control efforts, one that both enjoys wider international support and fits well with China's regional security interests. China also sees international disarmament and arms control forums as useful platforms for projecting its image as a developing country that has nuclear weapons but shares international concerns regarding nuclear arms control, thereby underlining its position as a Third World leader.

Against this background, China is seen as likely to continue to be cautious in dealing with questions related to nuclear test bans. On one hand, Beijing is inclined to welcome a U.S.-Soviet test ban agreement that would slow the superpower arms race, and it has indicated its willingness to participate in some discussions on the subject of banning nuclear tests.[18] Indeed, Beijing is seen as concerned that continued rapid advances in Soviet arms development may leave China further behind the Soviet Union, to a point where China's ability to deter Soviet intimidation may be called into question at some point in the future.

On the other hand, Chinese and U.S. observers note that such a test ban might increase unfavorable international attention to China's continued nuclear testing. They also note that a ban could result in pressure on China to join this kind of an arms control regime.

The Chinese government, therefore, has made it clear that it is not prepared to join a Comprehensive Test Ban agreement unless both Washington and Moscow agree to undertake what Beijing calls "deep cuts" in their nuclear arsenals.[19] At the least, China wants to be sure that a CTBT regime does not permanently leave the country in a seriously disadvantageous position against its likely adversary. The "deep cuts" are presumably a means to this end.

Although the Chinese government is not known to have taken a position on a quota test ban agreement, Chinese and U.S. observers foresee little problem for Beijing in this area. This is because China is not seen as likely to test nuclear weapons more than a few times a year.

A threshold test ban agreement at some lower yield than 150 kt could cause more problems for China. Chinese planners are said to be striving to develop smaller warheads for their missiles and bombers.[20] But China's techniques for nuclear testing are thought to be less advanced than those of the United States and the Soviet Union. Therefore, China might need to test at a higher threshold than the United States and Soviet Union might ultimately find acceptable.

### Nonnuclear Allies' Views Toward a U.S.-Soviet Comprehensive Test Ban

America's non-nuclear European allies generally favor U.S.-Soviet progress toward a CTBT as a means to prevent escalation of the arms race. But the subject of nuclear testing has not been a particularly divisive issue between the United States and its NATO allies, primarily because it is something over which the Europeans are split themselves. As already noted, under current circumstances Britain and France want continued testing to support their own nuclear weapons programs. The Federal Republic of Germany and other Alliance members would like to see progress made towards a gradual solution of the testing issue in the context of far-reaching cuts in offensive weapons.[21] German Foreign Minister Hans-Dietrich Genscher expressed hope that the U.S.-Soviet summit would lead to a "closer approximation of superpower views of the question of verification."[22] The German government, however, has been careful not to push the Reagan Administration too hard on the issue of testing. The negative reaction to President Reagan's Reykjavik proposal to eliminate nuclear ballistic missile in ten years demonstrated European sensitivity to any proposal that might undermine the credibility of the U.S. nuclear guarantee for Western Europe. West German support for a CTBT is therefore qualified by the concern that the reliability of the U.S. nuclear umbrella not be undermined. These fears would not necessarily be alleviated by any assurance that the Soviets also would stop testing. To the extent that a CTBT is perceived as lessening the risk that nuclear weapons might be used in a conflict--thereby increasing the importance of conventional forces--some Europeans would object to a U.S.-Soviet CTBT because of the NATO/Warsaw Pact imbalance in conventional arsenals.

To the extent that a CTBT is part of a broader nuclear and conventional arms reductions, however, most Europeans would favor a halt to testing. Some NATO countries such as Norway, Denmark, the Netherlands and Greece--strong supporters of a CTBT--have questioned the Administration's contention that inadequate verification measures are standing in the way of a test ban agreement. In contrast to the European nuclear powers, these nations would welcome a CTBT, perceiving such an agreement as a possible first step providing impetus to the superpowers to make significant nuclear arms control reductions.

### Some Implications of More Restrictive Test Bans for the Non-Proliferation Regime

A major argument in favor of a Comprehensive Test Ban Treaty (CTBT) is that it could bolster the worldwide regime that works against further spread, or proliferation, of nuclear weapons. Establishing a comprehensive test ban probably would be seen by many as a substantial new step to dissuade nonnuclear weapons states from opting for their own nuclear forces, especially the group of six states--Argentina, Brazil, India, Pakistan, Israel, and South Africa--now considered to be near or at the threshold of nuclear weaponry. In addition, a CTBT could dampen complaints of many nonnuclear weapons states about slow progress of the nuclear weapons states in negotiations for arms control as called for under the Treaty on the Non-Proliferation of Nuclear Weapons (NPT). This could be important as the year 1995 approaches when the extension of the NPT must be decided by an international conference and where the United States and the Soviet Union presumably will be seeking votes for extension. The longer that a CTBT is delayed, the greater the prospects that some important nonnuclear weapons NPT states may make it difficult for the

United States--and the Soviet Union, for that matter--to get an extension of the NPT on terms favorable to U.S. security interests. With this in mind, there follows a brief description of the international non-proliferation regime, and a discussion of the potential effects of a CTBT, both positive and negative, upon the regime.

## The International Non-Proliferation Regime and Its Parts

Ever since the United States dropped atomic bombs on Hiroshima and Nagasaki in 1945, the U.S. Government has worked to prevent the spread, or proliferation, of nuclear weapons. To that end, it led in erecting what is now commonly known as the world non-proliferation regime whose overall purpose is to reduce nuclear arsenals among the five nuclear weapons states--the United States, the Soviet Union, the United Kingdom, France and the People's Republic of China--and to prevent more states from getting nuclear weapons.[23] The regime, however, did not stop India from testing a "peaceful nuclear explosive" in 1974.

Today the international non-proliferation regime consists of a set of international commitments and obligations laid down in treaties, an international organization to verify the keeping of no-nuclear-weapons pledges, voluntary understandings of major nuclear supplier countries not to provide certain kinds of nuclear assistance and supplies to non-weapons states, a network of bilateral agreements for nuclear cooperation that incorporates some additional peaceful use commitments, and a pervasive world sentiment against further proliferation. A complementary set of U.S. trade, aid and diplomatic policies, established by legislative and executive actions, support and in some cases go beyond the regime.

The best-known part of the regime is the Treaty on Non-Proliferation of Nuclear Weapons (NPT) with its no-weapons pledges by nonnuclear weapons parties, pledges that are verified by national accounts of nuclear materials subject to international inspection (safeguards) to assure that these nuclear materials have not been diverted from peaceful nuclear activities to weapons.[24] This inspection responsibility is assigned to the International Atomic Energy Agency (IAEA).[25] The Latin American Nuclear Free Zone Treaty (Tlatelolco), the Partial-Test-Ban Treaty, the Threshold-Test-Ban and Peaceful Nuclear Explosion Treaties, and the new South-Pacific-Nuclear-Free-Zone Treaty are also parts, as would be a comprehensive test ban treaty if and when one is concluded.

The operating part of the regime is the IAEA which is assigned a verification function by the NPT. This verification is accomplished by international inspection and related measures "for the exclusive purpose of verification of the fulfillment of . . . obligations assumed under this Treaty with a view to preventing diversion of nuclear energy from peaceful uses to nuclear weapons or other nuclear explosive devices."[26]

### *Linkage of a CTBT with the Non-Proliferation Regime*

The Non-Proliferation Treaty has been described as the first operating nuclear arms control treaty. In turn, it refers to a comprehensive test ban directly in its preamble and indirectly in its body. The preamble declares an intention of parties to "achieve at the earliest possible date the cessation of the nuclear arms race and to undertake effective measures in the direction of nuclear disarmament." It urges cooperation among all states in attaining this objective and seeks the "discontinuance of all test explosions of nuclear weapons for all time." In Article VI of the NPT, parties are committed to "pursue negotiations in good faith on effective measures relating to cessation of the nuclear arms race at an early date and to nuclear disarmament, and on a treaty on general and complete disarmament under strict and effective international control."

The NPT does not prohibit nonnuclear weapons members from producing uranium-235 or plutonium-239, which can also be used to make nuclear weapons. Rather, the NPT prohibits them from acquiring or making nuclear weapons or nuclear explosives. An NPT non-weapons state therefore is free to produce and store weapons-grade materials for peaceful purposes as long as it permits IAEA inspection to verify that these can be accounted for.

A similar situation exists for parties to the Treaty of Tlatelolco, with two notable differences. Tlatelolco prohibits the stationing of nuclear weapons within the territories of its Latin American members, while the NPT does not. Thus, U.S. nuclear weapons can be found in European states that are parties to the NPT. Also Tlatelolco would permit acquisition of peaceful nuclear explosives under certain ambiguous conditions whereas the NPT bans both acquisition of peaceful explosives and atomic weapons by non-weapons parties. A recent addition to the regime is the South Pacific Nuclear Free Zone, or the Treaty of Rarotonga, which bans possession of peaceful nuclear explosives and nuclear weapons by its parties and also the stationing of nuclear weapons by any states within the territories of its parties.[27]

***The Non-Aligned Bloc and Its Importance.*** Over recent years the non-aligned bloc of nations, often referred to as the Group of 77, has complained that their faithful abstinence from nuclear weapons under the NPT has not been matched by corresponding progress of the nuclear weapons states towards nuclear disarmament.[28]

The Group of 77 now numbers about 100 developing states of the third world, most of them from Africa, Asia and Latin America. It was organized in the 1960s to deal with questions of trade with the industrial nations.[29] Since then many of its parties have become members and supporters of the non-proliferation regime and advocates of nuclear arms control. Of 99 members in 1986, 56 had ratified the Limited Test Ban Treaty and 11 more had signed but not ratified; 12 had ratified the Treaty of Tlatelolco, with one signing only; and 66 had ratified the NPT, with one signing only.

The importance of the Group of 77 is worthy of note. Within the decade they will have notable leverage to push for more progress in arms control, for they constitute more than a majority of the parties to the NPT. Looking ahead, in 1995 a fateful conference will be held to decide whether the Treaty shall "continue in force indefinitely, or shall be extended for an additional fixed period or periods." This decision will be by a majority vote of the parties.[30] So in the early 1990s the Group of 77 can be expected to bargain strongly for their support for extension of the NPT on terms favorable to U.S. interests and security. It is prudent to expect many non-aligned NPT states will be reluctant to support its extension without substantial U.S. concessions. In addition, they are likely to want substantial evidence of progress towards nuclear disarmament. Many of them see the NPT as a means for the nuclear weapons states to continue to expand their arsenals while the non-weapons states remain tightly bound by their non-proliferation pledges and are unable to attain whatever benefits nuclear weapons status might provide. So the non-aligned nations can be expected to argue for a comprehensive test ban as an important step toward ending the nuclear arms race and also as the most feasible and readily attainable measure to halt the further spread of nuclear weapons (horizontal proliferation) and further increases in nuclear arsenals (vertical proliferation).

As noted by a Canadian arms control expert, William Epstein,[31] the G-77 argues that if weapons testing is forbidden, the nonnuclear weapons countries would be less likely to try to make nuclear weapons, and the nuclear weapons states could not develop new weapons.[32] This linkage between non-proliferation pledges and nuclear arms control was much in evidence at the quinquennial review conferences for the NPT in 1975, 1980 and 1985. A sample of typical statements appears in Appendix C.

*The Group of Six and Its Importance.* A smaller group of third world nations remains adamantly opposed to non-proliferation pledges of the NPT or Tlatelolco. These include Argentina and Brazil, India and Pakistan, Israel and South Africa. Each of the six has caused concern in Congress during the past decade because of their continuing efforts to acquire the wherewithal to produce high quality nuclear materials suitable for weapons use--namely uranium-235 and plutonium. Pakistan's nuclear activities were much on the mind of Congress as it worked on extension of U.S. military and economic aid to that country. Each of the others has attracted the suspicion of various analysts and policymakers who see them at, or approaching, the threshold of a nuclear arsenal. Members of the group of six can be expected to argue against U.S. efforts to strengthen the non-proliferation regime and perhaps to discount the benefits of a test ban treaty. Nonetheless, several of them have made various kinds of no-weapons pledges in the Antarctic Treaty, the Partial Test Ban Treaty, the Outer Space Treaty, the Treaty of Tlatelolco and the Sea Bed Treaty as shown in Table 10. A cautionary note: The six include some with shaky governments and some in unstable parts of the world. So it cannot be assumed that their current policies on nuclear weapons will remain unchanged in 1995.

### Potential Effects of More Restrictive Test Bans

In general, the effects of a particular kind of test ban upon the non-proliferation regime and U.S. policy will depend upon how complete the ban may be. For example, would an agreement ban all nuclear weapons and nuclear explosive testing? Would all five nuclear-weapons states agree to it? How many and which of the nonnuclear-weapons states would become parties? Would it permit a quota of tests for the nuclear weapons states or permit them to test small devices whose yield is below a reduced threshold level? The more complete the ban, the more it would be expected to strengthen the non-proliferation regime and world support for it and U.S. non-proliferation policies. These benefits, however, would have a limited lifetime if not accompanied by other steps towards nuclear disarmament. Test bans that permit nuclear weapons states to continue testing under a quota or under a lowered threshold, would be less welcome by the non-weapons states and could evoke cynicism that would weaken their support for the regime, for extension of the NPT in 1995, and for U.S. non-proliferation policy.

Many potential positive and negative effects of a comprehensive test ban treaty or a limited or qualified test ban

**Table 10**

**Membership of the Group of Six
in Six Arms Control Treaties
(Years of Ratification or Accession)**

| State | Antarctic Treaty | Partial Test Ban Treaty | Outer Space Treaty | Treaty of Tlatelolco | Non-Proliferation Treaty | Sea-bed Treaty |
|---|---|---|---|---|---|---|
| Argentina | 1961 | 1986 | 1969 | * | | + |
| Brazil | 1975 | 1964 | 1969 | 1968 ** | | 1988 |
| India | 1983 | 1963 | 1982 | | | 1973 |
| Israel | | 1964 | 1977 | | | |
| Pakistan | | 1988 | 1968 | | | |
| South Africa | 1960 | 1963 | 1968 | | | 1973 |

\*   Argentina signed the treaty in 1967 but has not ratified it.
\*\*  Brazil has ratified the treaty subject to conditions not yet fulfilled.
\+   Argentina signed the treaty in 1971 but has not ratified it.

Source:  U.S. Arms Control and Disarmament Agency:  *Arms Control and Disarmament Agreements: Texts and
Histories of the Negotiations*, 1990 edition.  Washington, U.S. Govt. Print. Off., p. 29, 48-50, 60-62, 87,
103-106, and 115-117.

can be anticipated.    Some of the more plausible effects are summarized below.

*Positive Effects of a CTBT Cited by Supporters.*    A comprehensive test ban treaty could be expected to:

- Sustain and increase world support for the non-proliferation regime and reinforce the world opposition to more nuclear weapons; and enhance the standing of the NPT as a foundation for nuclear arms control.
- Dissuade threshold nuclear-weapons states--Argentina, Brazil, India, Israel, Pakistan and South Africa--from further developing nuclear weapons or the capacity to make them.
- Attract participation of states that will not join the NPT because they see it as discriminating in favor of the nuclear weapons states.    Ratification of a CTBT by France and by China, which are not NPT states, could make a CTBT more attractive.
- Meet demands of NPT nonnuclear-weapons for progress in nuclear arms control under NPT article VI.    These states have persistently pressed for such progress at the three NPT review conferences.
- Improve the prospects that the NPT will be extended in 1995 on terms favorable to U.S. non-proliferation interests.
- Influence Argentina and Brazil to become full parties to the Latin American Nuclear Free Zone Treaty (Tlatelolco).
- Strengthen reasons for nuclear supplier states to continue restrictions on supply of sensitive nuclear technologies to nonnuclear-weapons states, and strengthen U.S. support for such restrictions.
- Increase the importance of IAEA safeguards and their verification that civil nuclear activities are not diverted to the manufacture of nuclear weapons.
- Strengthen the influence of leaders within threshold states who oppose nuclear weapons.

*Negative Effects of a CTBT Cited by Opponents.* A comprehensive test ban treaty could be expected to:

- Leave unresolved the political and security problems that could cause some nations to want nuclear weapons.
- Not prevent a non-weapons state from making untested nuclear weapons.
- Raise questions about the reliability of the protection of the U.S. nuclear umbrella for its allies, and perhaps cause them to seek their own security by developing their own nuclear forces.
- Ultimately lose its effect unless accompanied by nuclear arms reductions.
- Encourage the creation of additional nuclear weapons free zones, some of which might jeopardize U.S. regional security interests, particularly in Western Europe.

*Positive Effects of a Reduced Threshold Test Ban or a Quota Treaty.* A reduced threshold test ban or a quota treaty could be expected to:

- Reassure NPT nonnuclear weapons states of some progress towards the arms control goal of NPT article VI.
- Reduce political controversy and bargaining as the NPT nears 1995 and the world conference for its extension.

*Negative Effects of a Reduced Threshold Test Ban or a Quota Treaty.* A reduced threshold test ban or a quota treaty could be expected to:

- Cause cynicism by nonnuclear weapons states which could erode their support of the regime and extension of the NPT in 1995.
- Cause some nonnuclear weapons states to decide nuclear arms control is unattainable and therefore to rethink their no-weapons pledges.
- Be seen as serving the special interests of the nuclear weapons states to the continuing disadvantage of non-

weapons states, thereby weakening their support for extension of the NPT in 1995.

## *Conclusion*

On the whole, it seems reasonable to expect that achieving a comprehensive test ban treaty would notably strengthen world support for the non-proliferation regime and for U.S. policies to prevent further proliferation, and improve prospects for extension of the NPT in 1995. On the other hand, this benefit would not be long lived if there was no additional progress towards nuclear arms control. As for a reduced threshold treaty or a quota treaty, either would do little to strengthen world support for the regime and for U.S. policy. Indeed, either might tend to weaken them. From this viewpoint, the net benefits of either from the nonproliferation perspective seem likely to be small.

## Notes

1.   Arms Control Reporter. Brookline, MA. 1988. p. 608.A.4. Includes announced tests plus those unannounced but detected by seismic means and announced by other public institutions as of Jan. 1, 1988.
2.   Interview with French government official.   November 14, 1986, Washington, D.C.
3.   Arms Control Reporter, 1988. p. 608.A.4.
4.   News from France. French Embassy Press and Information Service. Vol. 88.11.  June 14, 1988.
5.   Information from this section was collected through an interview with a ranking French government official who is a specialist in French security policy.  November 14, 1986.
6.   Oberdorfer, Don.  U.S. Rebuffs Nuclear Treaty for Pacific Nuclear Free Zone.  Washington Post, February 6, 1987: A-17.
7.   Arms Control Reporter, 1988. p. 608.A.4. Includes announced tests plus those unannounced but detected by seismic means and announced by other public institutions as of Jan. 1, 1988.  Does not include UK/US tests.
8.   Ibid. This figure excludes joint U.S./U.K. tests.
9.   De Young, Karen. British Confident They've Curbed U.S. Excesses. Washington Post, November 19, 1986: A-21.
10. U.K. Spells Out Its Policy on Arms Pact. Financial Times, November 5, 1986: 1.

11. White, David. UK deterrent would be reviewed if arms talks led to big cuts. Financial Times. May 19, 1988.

12. NATO Report, September 13, 1985: 6.

13. Statement on the Defence Estimates 1988. HMSO, London, England. Vol. I, p. 11.

14. Washington Post, August 20, 1986: A-24.

15. Jones, Michael and John Witherow. Thatcher to Fly to Reagan Talks. London Sunday Times, October 5, 1986: 1.

16. Arms Control Reporter, 1988. p. 608.A.4.

17. Consultations, Washington, D.C., November 1986, March 1988. For background, see: Johnston, Alastair I. China and Arms Control. Ottawa, The Canadian Center for Arms Control and Disarmament, 1986. 85 p. Jencks, Harlan. From Muskets to Missiles. Boulder, Colorado. Westview Press, 1983; and U.S. Congress. Joint Economic Committee. China's Economy Looks Toward the Year 2000. Chinese Nuclear Weapons and American Interests by Robert Sutter. 99th Cong., 2d Sess., v. 2. Washington, U.S. Govt. Print. Off., May 21, 1986. p. 169-185.

18. China has said it would participate in such discussions at the Geneva-based Committee on Disarmament. See, for instance, Xinhua, February 13, 1986, replayed in FBIS Daily Report, China, February 18, 1986.

19. See, for instance, Address of Chinese Representative at a U.N. Arms Control Forum, replayed in Beijing Review, March 30, 1987.

20. See Jencks and Harlan, PRC Nuclear and Space Programs. SCPS Yearbook on PLA Affairs, 1987. Sun Yat-sen Center for Policy Studies, National Sun Yat-sen University, Kaohsiung, Taiwan, R.O.C.

21. Statements and Speeches. Vol. IX, no. 19. German Information Center, November 7, 1986: 5.

22. Washington Post, August 20, 1986: A-21.

23. While a nuclear weapon and a nuclear explosive device intended for peaceful use both employ the enormous energy release and explosive force from rapidly fissioning atoms of uranium-235 or plutonium, there are notable differences. A nuclear weapon needs to be self contained, reliable in use under field conditions and predictable in yield. On the other hand, for certain underground peaceful uses sophisticated designs may be required to achieve a compact size and reduce residual radioactive contamination.

24. More details about the non-proliferation regime appear in Appendix A.

25. Appendix B contains a discussion of the usefulness of IAEA safeguard systems and technology for arms control verification.

26. Treaty on the Non-Proliferation of Nuclear Weapons, Article III-1.

27. The Treaty of Rarotonga entered into force on December 11, 1986. See pp. 210-211 for a brief description of the South Pacific Nuclear Free Zone Treaty.

28. Article VI of the NPT commits the United States, the Soviet Union and the United Kingdom to "pursue negotiations in good faith on effective measures relating to cessation of the nuclear arms race at an early date and to nuclear disarmament, and on a treaty on general and complete disarmament under strict and effective international control."

29. The members of the G-77 and their status with respect to the Limited Test Ban Treaty of 1963, the Treaty of Tlatelolco and the NPT are listed in Appendix D.

30. The NPT, Article X, paragraph 2.

31. Dr. William Epstein was former director of Disarmament in the U.N. Secretariat, a Senior Research Associate at Carleton University in Ottawa, and a member of the Canadian delegations to several sessions of the U.N. General Assembly. He represented the U.N. Secretary General at negotiations leading to the 1963 Test Ban Treaty, the 1967 Treaty of Tlatelolco, and the NPT.

32. William Epstein. A critical time for nuclear nonproliferation. Scientific American, August 1985: 35.

# Appendix A

## The International Nuclear Non-Proliferation Regime

*Warren H. Donnelly*

The International Nuclear Non-Proliferation Regime, as noted earlier, consists of a set of treaties, an international organization to verify the no-nuclear-weapons pledges of non-weapons states belonging to the non-proliferation treaty, voluntary agreements by nuclear supplier states to require certain conditions for their nuclear exports and to exercise restraint in supply of sensitive nuclear technologies and items, together with a worldwide predisposition against further spread of nuclear weapons and a desire in many, although not all, quarters for substantial nuclear disarmament. The treaties, the international agency and the voluntary commitments will now be described.

### The Non-Proliferation Treaty

The heart of the regime is the Treaty on the Non-Proliferation of Nuclear Weapons (NPT) which took effect in 1970. In essence, it froze the number of nuclear weapons states at five and divided nations into two groups: those that possessed nuclear weapons at that time, and those that did not.

Each NPT nuclear weapons state agrees not to transfer to any recipient whatsoever nuclear weapons or explosive devices directly or indirectly; and not in any way to assist, encourage, or induce any nonnuclear-weapon state to manufacture or otherwise acquire nuclear weapons or other nuclear explosive devices, or control over such weapons or explosive devices.[1] France and China have not joined the NPT. France, however, has said it would behave as though it was a party.

For their part, each nonnuclear-weapon member state "...undertakes not to receive the transfer from any transferor whatsoever of nuclear weapons or other nuclear explosive devices or of control over such weapons or explosive devices directly, or indirectly; not to manufacture or otherwise acquire nuclear weapons or other nuclear explosive devices; and not to seek or receive any assistance in the manufacture of nuclear weapons or other nuclear explosive devices."[2] They are not required, however, to ban the stationing of nuclear weapons of other states within their territories. Note, however, that stationing of nuclear weapons in non-weapons states is barred by the Latin American Nuclear Free Zone Treaty and the South Pacific Nuclear Free Zone Treaty.

Additionally, each nonnuclear-weapon state undertakes to accept international inspection (safeguards) by the International Atomic Energy Agency for the "exclusive purpose of verification of the fulfillment of its obligations under the Treaty" with a view to ". . . preventing diversion of nuclear energy from peaceful uses to nuclear weapons or other nuclear explosive devices."[3] These safeguards apply on all source or special fissionable material in all peaceful nuclear activities within a non-weapons state, under its jurisdiction, or carried out under its control anywhere.[4] The three NPT nuclear weapons states--the United States, the United Kingdom and the Soviet Union--have voluntarily opened some of their civil nuclear facilities to IAEA inspection. France has done likewise and China is completing negotiations with the IAEA for such inspection.

All NPT states further are committed not to provide nuclear materials, or equipment, or material especially designed or prepared for the processing, use, or production of special fissionable material, to any nonnuclear-weapon state unless these items are subject to IAEA safeguards.[5] France and China also have made this a condition for their nuclear exports. Note, Article IV in effect permits non-weapons members to produce and stockpile weapons grade uranium and plutonium without limits for civil purposes, and specifies that nothing in the treaty shall affect the "inalienable right" of all parties to "develop research production, and use of nuclear energy for peaceful purposes without discrimination. . . ."

Finally, the nuclear weapons states are committed to negotiate "effective measures relating to cessation of the nuclear arms race at an early date and to nuclear disarmament, and on a treaty on general and complete disarmament under strict and effective international control."[6]

By November 1987, 133 nonnuclear weapons states had ratified the NPT and taken its pledge not to acquire nuclear weapons, and 37 states with notable nuclear activities had concluded safeguards agreements to permit IAEA inspection. However, the continued refusal of 12 nonnuclear weapons states, some with advanced nuclear programs, to join the NPT has diminished its effectiveness. Among these states are several with notable nuclear programs: Argentina, Brazil, India, Israel, Pakistan and South Africa.

## The Treaty of Tlatelolco

A notable, although still incomplete, part of the regime is the Treaty for the Prohibition of Nuclear Weapons in Latin America, known as the Treaty of Tlatelolco.[7] It is the first treaty to establish a nuclear weapons free zone in any populated part of the world and predates the NPT. Although it is not yet in effect for Argentina, Brazil, Chile, and Cuba, other Latin American states that have ratified it have exercised their option to bind themselves to it before all the requirements are met.

The treaty prohibits indigenous development of nuclear weapons by member states and the testing of nuclear weapons in their territories, including underground testing. Further, it goes beyond the NPT by prohibiting the receipt, storage, installation, or deployment of nuclear weapons in their territories. Thus the treaty obliges its parties not to allow any outside nation to install, store, or deploy nuclear weapons within its territories. A protocol signed by all five nuclear weapons states commits them to respect the Treaty and, in addition, not to use or threaten to use nuclear weapons against its parties.

As for verification, the treaty requires parties to enter into safeguards agreements with the International Atomic Energy Agency (IAEA). In addition, it has established a regional control organization, OPANAL,[8] which has the right to obtain information from parties about their nuclear activities and to conduct special inspections.

## The Treaty of Rarotonga

A recent addition to the regime is the South Pacific Nuclear Free Zone Treaty, better known as the Treaty of Rarotonga, which entered into force on December 11, 1986. In addition to Australia, other countries party to the nuclear free zone include the Cook Islands, Fiji, Kiribati, Nauru, New Zealand, Niue, Papua New Guinea and the Solomon Islands, Tonga, Tuvalu, Vanuatu and Western Samoa. The treaty commits its parties: (1) not to manufacture or otherwise acquire, possess or control any nuclear explosive device by any means; (2) to prevent the stationing of any nuclear explosive device in their territories; (3) to prevent the testing in their territories of any nuclear explosive device; and (4) not to assist or encourage the testing of any nuclear explosive device by any state. However, each Party "... remains free to decide for itself whether to allow visits by foreign ships and aircraft to its ports and airfields, transit of its airspace by foreign aircraft, and navigation by foreign ships in its territorial sea ... in a manner not covered by the rights of innocent passage ..." With its emphasis on banning all nuclear explosive devices, whether peaceful or military, the treaty creates a "nuclear free" zone rather than the "nuclear-weapons-free" zone of the Treaty of Tlatelolco. Three protocols to the Treaty of Rarotonga involve states external to the region. Protocol 1 would commit foreign states that have jurisdiction over territories within the zone to apply the Treaty's key provisions to these territories. Protocol 2 would commit the five nuclear weapons states not to use or threaten to use nuclear explosive devices against parties to the Treaty and Protocol 3 would commit the nuclear weapons states to refrain from testing within the zone. France, the United Kingdom and the United States all have yet to sign the protocols.

## The Partial Test Ban Treaty

An early treaty important to the regime is the partial test ban treaty (PTBT) which took effect in 1963.[9] The parties to this treaty undertake "not to carry out any nuclear weapon test explosion, or any other nuclear explosion," in the air, under water or in outer space. As explained by the State Department during the Kennedy Administration, the phrase "any other nuclear explosion" includes explosions for peaceful purposes. These are prohibited by the treaty because of the difficulty in differentiating between them and nuclear weapons tests.[10] The treaty was negotiated by the three original nuclear weapons states--the United States, the United Kingdom and the Soviet Union--and in 1985 had 112 signatories. China and France are not parties. Ratification of the treaty by many nonnuclear weapons states is seen as evidence of their support for non-proliferation measures.

## The Threshold Test Ban and the
## Peaceful Nuclear Explosions Treaties

Signed with the Soviet Union but not yet ratified by the United States, the Threshold Test Ban Treaty (TTBT) and the Peaceful Nuclear Explosions Treaty (PNET) could become outlying parts of the nonproliferation regime and a precursor to or a substitute for a full comprehensive test ban treaty.

The TTBT was signed by the United States and the Soviet Union in 1974. It would ban underground nuclear weapons tests with a yield of more than 150 kilotons (kt), and provide for an exchange of data on calibration of nuclear test yields in order to assist verification. The Nixon Administration did not seek ratification of the TTBT pending negotiation of a companion treaty to deal with peaceful nuclear explosives.

The PNET was signed by the superpowers in 1976. It would ban individual nuclear explosions above 150 kilotons (kt) yield, or any group explosion with an aggregate yield of 1,500 kt, for peaceful purposes and would provide for on-site inspection in certain circumstances. Notable is the requirement for special verification procedures when the aggregate yield of a group of explosions is more than 150 kt. In this case, the verifying side would have the right to have observers and instruments at the site of a group explosion to determine the yield of each device. In addition, observers could be permitted on the basis of consultations between the parties explosions with an aggregate yield between 100 and 150 kt.[11] The Director of the Arms Control and Disarmament Agency at the time, Mr. Paul C. Warnke, commented on the verification provisions for these two treaties to the Senate Committee on Foreign Relations at hearings in 1977 as follows:[12]

> . . . And several precedents will be important and relevant. The recognition that each side must furnish data to assist the other's national technical means of verification is significant. Of even more significance, moreover, is the recognition that in some cases even national technical means in combination with the data furnished should be supplemented to ensure adequate verification. Where this occurs, the principle will be established by the PNE Treaty that observers with equipment are authorized to assure compliance with treaty provisions. On-site verification is a valid tool that can be helpful in establishing an adequate verification regime in a comprehensive ban on all nuclear tests including explosions for peaceful purposes.

The two treaties were submitted to the Senate for ratification by President Ford in 1976, where they still remain on the calendar of the Foreign Relations Committee.

The Reagan Administration contended that TTBT and PNET verification provisions were inadequate, that the Soviet Union probably tested at yields above 150 kilotons, and that testing is needed as long as deterrence requires nuclear weapons. It therefore broke off CTBT negotiations in June 1982, and indicated that the United States would not ratify the TTBT or PNET unless they were renegotiated to improve their verifiability. In February 1983, the United States asked the Soviet Union to open talks on improving verification. The Soviet Union declined, arguing that the treaties were adequately verifiable and that problems verifying them would not arise under a CTBT. To show its desire for a CTBT, the Soviet Union held a moratorium on its own tests from August 1985 to February 1987, which the United States declined to join.

In May 1986, the House amended the FY87 defense authorization bill to call for a U.S.-Soviet moratorium on nuclear tests above 1 kt. The Administration fought this amendment. In a compromise, conferees on the bill dropped it. In return, the President agreed to inform General Secretary Gorbachev at the Reykjavik summit of October 1986 that he would (1) proceed promptly with ratification of the TTBT and PNET if verification concerns were met, and (2) would then proceed with negotiations on step-by-step measures to further limit and ultimately end nuclear testing in parallel with measures to reduce and ultimately eliminate nuclear weapons. At the summit, the leaders agreed to this approach.

Negotiations are proceeding on TTBT and PNET verification. At the Moscow summit of May 1988, Reagan and Gorbachev signed a Joint Verification Experiment Agreement to improve the verifiability of the TTBT and PNET. A U.S. team will visit the Soviet test site at Semipalatinsk, and a Soviet team will visit the U.S. test site in Nevada. Each host nation will conduct a nuclear explosion, and both sides will perform experiments on means of measuring the yield.

## The Nuclear Suppliers Guidelines

Another notable part of the non-proliferation regime is the set of voluntary guidelines for nuclear exports agreed upon by the Nuclear Suppliers Group, popularly known as the London Club. Because of the dangers of spreading the ability to make nuclear weapons materials through uncontrolled sale of sensitive nuclear technology and equipment,[13] seven major nuclear suppliers met in London in 1975 to establish common guidelines for their nuclear exports.[14] Eight other suppliers later joined.[15] The suppliers guidelines call for restraint in the transfer of sensitive nuclear technology, equipment and materials, and specify special restrictions for the export of items on a "trigger list."[16] The conditions of supply include assurances of non-explosive use by the recipient, effective physical protection of nuclear materials, and IAEA safeguards for the supplied items. The guidelines were agreed to in secret in 1976. Two years later they were formally communicated to the International Atomic Energy Agency and made public.

While the guidelines in effect are an international gentlemen's agreement and members can withdraw their support at any time, through the time of writing there have been no reports of deliberate evasions of them by members.

## Notes

1. NPT Article I.
2. NPT Article II.
3. NPT Article III.
4. Source materials are natural uranium and its chemical compounds; special nuclear materials are primarily uranium enriched in the U-235 isotope and plutonium. These terms are defined in section 11 of the Atomic Energy Act of 1954 as amended.
5. NPT Article III.
6. NPT Article VI.
7. Named after the suburb of Mexico City where that treaty was negotiated.

8.   The Agency for the Prohibition of Nuclear Weapons in Latin America, has its headquarters in Mexico City.   The Agency is responsible for holding periodic or extraordinary consultations among Member States on matters relating to the purposes, measures and procedures of the Treaty and for the supervision of compliance with the obligations arising therefrom.

9.   The Treaty Banning Nuclear Weapons Tests in the Atmosphere, in Outer Space and Under Water.

10.   U.S.   Arms Control and Disarmament Agency.   Arms Control and Disarmament Agreements.   Texts and Histories of Negotiations.   Washington, 1982 edition.   p. 40.

11.   At the time, this idea of on-site inspection was thought quite radical. However, now that the INF Treaty has been ratified, its on-site inspection provisions make these provisions of the PNET seem rather mild.

12.   U.S.   Congress.   Senate.   Committee on Foreign Relations. Subcommittee on Arms Control, Oceans and International Environment. Threshold Test Ban and Peaceful Nuclear Explosion Treaties. Hearings, 95th Cong., 1st Sess., July 28, August 3 and September 8 and 15, 1977.  p. 8.

13.   For purposes of U.S. policy, Congress defined "sensitive nuclear technologies" in section 4 of the Nuclear Non-Proliferation Act of 1978 to mean any information which is not available to the public and which is important to the design, construction, fabrication, operation or maintenance of a uranium enrichment or nuclear fuel reprocessing facility or a facility for the production of heavy water.

14.   The initial members were Canada, France, West Germany, Japan, the United Kingdom, the United States and the Soviet Union.

15.   The eight additional included Belgium, Czechoslovakia, East Germany, Italy, the Netherlands, Poland, Sweden and Switzerland.

16.   The trigger list includes source and special fissionable materials, nuclear reactors and certain reactor components; non-nuclear materials for reactors, notably heavy water and nuclear grade graphite; reprocessing plants and associated equipment, fuel fabrication plants, heavy water plants, and equipment for separation of uranium isotopes.

# Appendix B

## International Verification: The IAEA

*Warren H. Donnelly*

Unique to the international non-proliferation regime is verification of no-nuclear-weapons pledges by international inspection. As noted earlier, non-weapons NPT states are committed to accept inspection by the International Atomic Energy Agency (safeguards) to verify their no-weapons pledges under the NPT with a view to preventing the diversion of nuclear energy from peaceful uses to nuclear weapons or other nuclear explosive devices. These safeguards are to be applied on all source material or special fissionable material in all peaceful nuclear activities under the jurisdiction or control of non-weapons member states.

The IAEA was established in 1957 by an international statute and is located in Vienna, Austria. Before the NPT took effect in 1970, the Agency had developed a safeguards system to be applied to some, but not necessarily all, nuclear facilities in states which voluntarily requested its application. U.S. policy in the 1960s was to require IAEA safeguards for its nuclear exports, thus providing working experience for the Agency. The NPT changed this spotty application of safeguards by requiring all non-weapons parties to negotiate safeguards agreements with the Agency to apply to all of their peaceful nuclear activities. Subsequently, four of the nuclear weapons states voluntarily negotiated agreements with the Agency to open some of their civil nuclear facilities to inspection.[1]

The keystone of the Agency's NPT safeguards system is its safeguards agreements with participating states which require them to establish and maintain national systems of accounting for and control of nuclear materials within their territories, and that spell out the agency's inspection rights. The focus of IAEA safeguards is verification. As explained by the Agency:[2]

> To verify means "to establish the truth of." In safeguards, to verify is to establish the truth of statements regarding the amounts, presence and use of nuclear material or other items subject to safeguards as recorded by facility operators and as reported by the State to the IAEA. Accountancy, taken together with containment and surveillance, is the fundamental basis on which verification rests.

IAEA verification is a three stage process which includes:

- Examination of the information provided by a state;
- Collection of information by the Agency as a result of inspections to verify information on design of nuclear facilities, inspections to examine records and reports, regular inspection of nuclear material, and special inspections; and
- Evaluation of the information provided by a State and of that collected by inspectors, to determine the completeness, accuracy and validity of the information provided by the State.

The purpose of nuclear materials accounting is to establish the quantities of nuclear materials present within defined times. Containment and surveillance help to reduce the probability that undetected movements of nuclear materials or equipment can take place. "Containment" puts barriers in the way of unauthorized access; "surveillance" provides a record that unauthorized access did not occur between inspections. These measures may involve the use of devices such as tamper-indicating seals, TV monitors and cameras, and observation posts.

IAEA verification concentrates on parts of the nuclear fuel cycle where weapons-usable materials--namely uranium-235 and plutonium-239--can be found.[3] These places include factories, storage facilities, and transportation terminals. Verification must be sufficient to enable the Agency to meet the defined objective of safeguards, which is the timely detection of the diversion of a significant quantity of nuclear material.[4]

As for verification technologies, the trend is to rely increasingly upon improved instrumentation, monitoring and surveillance systems, and other technological means to show that nuclear materials have not been diverted from civilian uses. Continuing expansion of the world's nuclear power generation, although slower than anticipated a few years ago, means more nuclear power plants. This expansion continues to increase the demands upon IAEA inspection resources, which tend to be limited by political considerations and an unwillingness by major IAEA members, including the U.S., to increase the Agency's budget.

## Notes

1. Article III of the NPT commits each non-nuclear weapons party to accept safeguards, as set forth in an agreement between the International Atomic Energy Agency and the country, for the exclusive purpose of "verification of the fulfillment of its obligations . . . with a view to preventing diversion of nuclear energy from peaceful uses to nuclear weapons or other nuclear explosive devices." These safeguards apply to "all source or special fissionable material in all peaceful nuclear activities" within the territory of the State, under its jurisdiction, or under its control anywhere. Note, source materials refer to natural uranium and its compounds; special fissionable materials refer to uranium enriched in the U-235 isotope or plutonium.

2. IAEA Safeguards, An Introduction. Vienna, Austria: International Atomic Energy Agency, 1981 (IAEA report SG/INF/3).

3. The term "nuclear fuel cycle" refers collectively to the various industrial operations needed to produce, fabricate, use and dispose of nuclear fuels, beginning with uranium mining and ending with disposal of used, or "spent," nuclear fuels.

4. IAEA safeguards, op.cit., passim.

# Appendix C

## Linkage of a CTBT with the Non-Proliferation Regime: Views from the NPT Review Conferences of 1975, 1980 and 1985

*Warren H. Donnelly*

The Non-Proliferation Treaty (NPT) called for a conference in 1975 to review performance of the treaty. Following that first conference, the United Nations passed resolutions that led to additional conferences in 1980 and 1985. Another is scheduled in 1990. Then in 1995 will come the conference to decide on extension of the treaty.[1] These quinquennial review conferences have provided a stage for many non-weapons NPT states to strongly press for more progress on nuclear arms control and sometimes to link the future of the international non-proliferation regime to the outcome of these arms control efforts. This appendix samples some of the more notable views at the conferences in 1975, 1980 and 1985.

### The 1975 NPT Review Conference

At the 1975 conference, the Group of 77 pushed the nuclear weapons states to: (1) end underground nuclear tests; (2) substantially reduce their nuclear arsenals; and (3) pledge not to use or threaten to use nuclear weapons against nonnuclear weapons parties to the NPT.[2] Tensions and disagreement between the Group of 77 and the nuclear weapons states almost prevented agreement upon a final declaration for the conference. Nonetheless, although the declaration barely papered over disagreements between the weapons and some non-weapons states, it did affirm the statements in the 1963 Partial Test Ban Treaty and in the preamble of the NPT supporting the discontinuance of all nuclear weapons tests. The declaration said that "a treaty banning all nuclear weapons tests is one of the most important measures to halt the nuclear arms race" and appealed to the nuclear weapons states to "make every effort to reach agreement on the conclusion of an effective nuclear comprehensive test ban."

## The 1980 NPT Review Conference

Five years later, in 1980, the Group of 77 nations argued that nuclear weapons states not only had ignored the 1975 declaration but had failed to implement the nuclear disarmament provisions of Article VI of the NPT. The group wanted the declaration from the 1980 conference to note this lack of arms control progress and, in particular, to express regret that a comprehensive test ban treaty had not been concluded and that multilateral negotiations to this end had not begun in the Committee on Disarmament in Geneva.[3] Going further, the G-77 bloc wanted a reaffirmation of the nuclear disarmament commitment of NPT Article VI and also a recommendation for the start of multilateral negotiations for a comprehensive test ban treaty and a moratorium by the nuclear weapons states on further testing pending conclusion of the test ban. The United States and Soviet Union urged adoption of a compromise declaration. But the Group of 77 would not agree to language that "merely papered over the differences" in order to reach another "phony consensus," as had happened, in their view, in 1975.[4]

The close linkage of weapons testing and non-proliferation was underscored by Dr. Sigvaard Eklund, then director general of the International Atomic Energy Agency. Addressing the review conference in 1980, he said:

> I can only repeat what I said here five years ago in reference to a major commitment expressed in article VI and in the preamble to the Treaty, namely, the declared determination of the Parties to achieve the discontinuance of all test explosions of nuclear weapons for all time and to continue negotiations to this end. I said then that an effective treaty banning every kind of nuclear weapons test would be the most important single action that could be taken to strengthen and universalize the beneficial regime of non-proliferation of nuclear weapons. Events since 1975 have only served to strengthen this conviction.

This view was echoed by other delegates. For example, the Norwegian delegate said that a comprehensive test ban would constitute a "non-discriminating instrument of essential relevance to the promotion of the objectives of non-proliferation." The Japanese delegate spoke of the importance of a test ban, not only as a measure to curb the qualitative improvement of nuclear weapons, but also as a measure to prevent nuclear proliferation, and added that this importance had been widely stressed in the world community. The Australian delegate said that conclusion of a test ban would be a major demonstration by the nuclear weapons states of their intention to work for nuclear disarmament, would bring pressure to bear on other nuclear weapon states to cease nuclear testing, and would serve as an additional and important support for the international non-proliferation regime.

## The 1985 NPT Review Conference

As the third review conference drew near, there were mixed feelings for its prospects. The Reagan Administration was optimistic and sent representatives to many non-weapon states to solicit their support for the NPT and to defuse potentially explosive criticisms of the slow pace in arms control. Others, however, saw a bleaker outlook. For example, the Canadian arms control expert, William Epstein, wrote:[5]

... The frustration of the neutral and nonaligned countries appears to be turning into resentment and anger because they believe the nuclear powers have misled them. These nations will no doubt renew their long-standing demands for a comprehensive test ban, nuclear disarmament, assurances that nuclear weapons will not be used or held out as threats against nonnuclear countries and greater assistance in the peaceful use of nuclear energy.

He outlined various scenarios for the Group of 77 to press for a test ban and other steps of nuclear arms control.

Another arms control analyst, Jozef Goldblat of Sweden,[6] recalled that the chances of a successful outcome had appeared rather low, saying:[7]

... Many observers expected sharp polemics between the USA and the USSR on nuclear arms control issues, as well as harsh criticism by Third-World countries of the superpowers' failure to start the process of nuclear disarmament. All of this seemed to rule out an agreement on the future course of action with respect to the implementation of the NPT, and to make the breakdown of the Conference inevitable, with all the negative consequences which such an occurrence could entail for the continued operation of the Treaty.

And in April 1985, shortly before the conference, a symposium at the United Nations addressed the question of "Survival in the Nuclear Age."[8] Its concluding statement said that a comprehensive test-ban treaty was crucial for the success of efforts to halt and reverse the nuclear arms race and to prevent the expansion of existing arsenals and the spread of nuclear weapons to additional countries. Of the non-proliferation connection, it said:[9]

1985 will see the third review of the non-proliferation treaty. The failure by the nuclear-weapon states during the last 15 years to fulfill their obligation under the NPT relating to the cessation of the nuclear arms race and the continuation of vertical proliferation of nuclear weapons increase the danger of horizontal proliferation and endanger the entire non-proliferation regime. There is need for concrete and meaningful action now if the non-proliferation treaty--which we emphasize is a most important international agreement--is not to become another victim of the arms race.

As things turned out, none of these predictions came true. The United States and the U.S.S.R. limited themselves to restatements of their well known positions without directly assailing each other. The Group of 77, although anxious for progress on arms control under Article VI and a test ban, delayed an expected frontal attack on the superpower arms control policies until it became clear that the United States was not prepared to move on a comprehensive test ban. In the controversial closing hours of the conference there were fears that it might collapse as the non-aligned states held out for endorsement in the final declaration of resumption of test ban negotiations. Eventually, and in the face of non-aligned threats to force a vote on three draft resolutions calling respectively for a moratorium on testing, resumption of negotiations for a comprehensive test ban, and a nuclear weapons freeze, the United States agreed to a compromise declaration which "deeply regretted" that a comprehensive test ban had not yet been concluded and called for the immediate resumptions of negotiations by the superpowers "as a matter of the highest priority."[10] The final document also noted that certain states (understood to be the United States and the United Kingdom) while committed to the goal of an effectively verifiable comprehensive nuclear test-ban, considered deep and verifiable reductions in existing arsenals as the highest priority in pursuit of the objectives set out in NPT article VI.

The 1985 review conference also heard many statements linking nuclear arms control, a test ban, and non-proliferation. The present IAEA director general, Dr. Hans Blix, spoke to the conference about the "prime importance" that all nations, and in particular, the great nuclear powers, should direct their policies to create and foster a political environment in which nonnuclear weapon states do not feel the need to have nuclear weapons or the capacity to make them. Elaborating, he said:[11]

> Nothing could do more to strengthen the barriers against further proliferation, however, than progress in nuclear arms control. If horizontal proliferation has been stemmed for the time being, we are all too well aware that the same cannot be said of its giant and immeasurable more dangerous brother, vertical proliferation.

A joint letter to the conference from the representatives of Denmark, Finland, Iceland, Norway and Sweden said that "The conclusion of a comprehensive test-ban treaty would effectively enhance the non-proliferation regime." Going further, it said:[12]

> . . . The Nordic Governments attach particular importance to the conclusion of a comprehensive test-ban treaty with universal adherence, which would be a most effective measure to halt further development of nuclear weapons and offer strong support for the purposes of the non-proliferation treat.

The Group of Non-Aligned and Neutral States in a working paper on Article VI of the NPT recalled and underscored a declaration from the Seventh Non-Aligned Summit of March 1983 which said, in part:

> . . . In order to prevent effectively the horizontal and vertical proliferation of nuclear weapons, nuclear weapons states should adopt urgent measures for halting and reversing the nuclear arms race.

As an interim measure, the non-aligned summit called for a nuclear weapons freeze and "the speedy finalization of a comprehensive Treaty banning the testing of nuclear weapons."[13]

Several post-conference writings from nonnuclear weapons states further illustrate thinking on this linkage. Jozef Goldblat of SIPRI wrote in January 1986 that a test ban would:[14]

> . . . place practical obstacles in the way of would-be proliferators, as governments may hesitate to build a significant stock of untested weapons. Moreover, since a comprehensive test-ban would apply to both nuclear-weapons and non-weapons states, it would partly obliterate the politically sensitive aspect of the Treaty--its implication that one group of states is permitted to develop and test nuclear weapons while another is not.

And the Canadian arms control expert, William Epstein, wrote in April 1986 that:[15]

> A total test ban would be a major, indeed indispensable, step to halt the nuclear arms race, by curbing the development of new and more destabilizing weapons by the present nuclear powers and by preventing the spread of nuclear weapons to nonnuclear nations.

If testing is banned, the nuclear powers would be unlikely to commit the vast resources required to develop nuclear weapons systems as the military could not be certain that any new weapon would work effectively without testing it. If the hopes of achieving superiority in the nuclear arms race were ended by stopping testing, there would be little or no incentive to continue that race and more reason to agree on reducing nuclear forces. For similar reasons the nonnuclear powers would also be unlikely to embark of a nuclear weapons program if they could not test their weapons.

Finally, a statement of the 36th Pugwash Conference on Science and World Affairs, which met in Budapest in September 1986, linked the test ban and non-proliferation in its concluding statement, saying:[16]

. . . security will be diminished on all sides if non-weapons states show their frustration at the failure of the weapons states, until now, to achieve a CTBT, by withdrawing from the Non-Proliferation Treaty, by acquiring nuclear weapons, or even by merely moving closer to acquiring them.

## Notes

1. Article VIII of the NPT provided for a conference to be held in 1975 to "review the operation of this Treaty with a view to assuring that the purposes of the Preamble and the provisions of the Treaty are realized." Thereafter at five year intervals, a majority of the parties could convene further conferences, which they have done. The last review conference will be held in 1990. Five years later, in 1995, the parties are to meet at a fateful conference to decide whether the treaty shall continue in force indefinitely, or shall be extended for an additional fixed period or periods.
2. William Epstein. Nuclear Proliferation: The Failure of the Review Conference. Survival, November/December 1975: 265.
3. William Epstein. On Second Review of the Non-Proliferation Treaty. The Bulletin of the Atomic Scientists, May 1981: 58.
4. Ibid., p. 59.
5. William Epstein. A Critical Time for Nuclear Non-Proliferation. Op. cit.: 37.
6. Dr. Jozef Goldblat is in charge of the Arms Control and Disarmament Program at the Stockholm International Peace Research Institute.
7. Jozef Goldblat. The Third Review Conference of the Nuclear Non-Proliferation Treaty. Bulletin of Peace Proposals, vol. 17, no. 1, 1986: 13.
8. The symposium was sponsored jointly by the Third World Foundation and Parliamentarians for World Order, under the chairmanship of Willy Brandt.

9. Survival in the Nuclear Age. Disarmament, Winter 1985: 153.

10. Final Document of the Third NPT Review Conference, Part I (NPT/Conf. III/64/I) Annex I, Article VI and preambular paragraphs 8-12, paragraph B.12, p. 14. See also William Epstein. Reviewing the Non-Proliferation Treaty, The Canadian Institute for International Peace and Security, Background Paper No. 4, Ottawa, March 1986: 4, 5; and NPT Only Given a Reprieve. Arms Control Today, November/December 1985: 2.

11. Hans Blix. Preventing the Spread of Nuclear Weapons: The International Atomic Energy Agency and the Trea Non-Proliferation of Nuclear Weapons. Geneva, 1985.

12. Review Conference of the Parties to the Treaty on the non-proliferation of nuclear weapons. Final document, Part II, Geneva, 1985, p. 2.

13. Ibid., p. 81.

14. Jozef Goldblat. Will the NPT Survive? Bulletin of the Atomic Scientists, January 1986: 38.

15. William Epstein. A Test Ban--Halting the Nuclear Arms Race. Christian Science Monitor, April 23, 1986: 14.

16. Pugwash on Common Security. Bulletin of the Atomic Scientists, November 1986: 48.

# Appendix D

## List of Non-Aligned Nations Showing Which Are Parties to the Limited Test Ban Treaty, the Latin American Nuclear Free Zone Treaty, and the Non-Proliferation Treaty

*Warren H. Donnelly*

| Country | Limited Test Ban Treaty | Treaty of Tlatelolco | Non-Proliferation Treaty |
|---|---|---|---|
| 1. Afghanistan | x | | x |
| 2. Algeria | * | | |
| 3. Angola | | | |
| 4. Argentina | * | * | |
| 5. Bahamas | x | x | x |
| 6. Bahrain | | | |
| 7. Bangladesh | | | x |
| 8. Barbados | | x | x |
| 9. Belize | | | |
| 10. Benin | x | | x |
| 11. Bhutan | x | | x |
| 12. Bolivia | x | x | x |
| 13. Botswana | x | | x |
| 14. Burkina Faso | | | x |
| 15. Burundi | * | | x |
| 16. Cameroon | * | | x |
| 17. Cape Verde | x | | x |

| Country | Limited Test Ban Treaty | Treaty of Tlatelolco | Non-Proliferation Treaty |
|---|---|---|---|
| 18. Central African Rep. | x | | x |
| 19. Chad | x | | x |
| 20. Ghana | x | | x |
| 21. Colombia | * | x | * |
| 22. Comoro | | | |
| 23. Congo | | | x |
| 24. Cuba | | | |
| 25. Cyprus | x | | x |
| 26. Djibouti | | | |
| 27. Ecuador | x | | x |
| 28. Egypt | x | | x |
| 29. Equatorial Guinea | | | x |
| 30. Ethiopia | * | | x |
| 31. Gabon | x | | x |
| 32. Gambia | x | | x |
| 33. Grenada | | x | x |
| 34. Guinea | | | |
| 35. Guinea Bissau | | | x |
| 36. Guyana | | | |
| 37. India | x | | |
| 38. Indonesia | x | | x |
| 39. Iran | x | | x |
| 40. Iraq | x | | x |
| 41. Ivory Coast | x | | x |
| 42. Jamaica | * | x | x |
| 43. Jordan | x | | x |
| 44. Kampuchea | | x | |
| 45. Kenya | x | | x |
| 46. Kuwait | x | | |
| 47. Laos | x | | x |
| 48. Lebanon | x | | x |
| 49. Lesotho | | | x |
| 50. Liberia | x | | x |
| 51. Libya | x | | x |
| 52. Madagascar | x | | x |
| 53. Malawi | x | | x |
| 54. Malaysia | x | | x |
| 55. Maldives | | | x |
| 56. Mali | * | | x |
| 57. Malta | x | | x |
| 58. Mauritania | x | | x |
| 59. Mauritius | x | | x |

| Country | Limited Test Ban Treaty | Treaty of Tlatelolco | Non-Proliferation Treaty |
|---|---|---|---|
| 60. Morocco | x | | x |
| 61. Mozambique | | | |
| 62. Nepal | x | | x |
| 63. Nicaragua | x | x | x |
| 64. Niger | x | | |
| 65. Nigeria | x | | x |
| 66. Oman | | | |
| 67. Pakistan | * | | |
| 68. Panama | x | x | x |
| 69. Peru | x | x | x |
| 70. Qatar | | | |
| 71. Rwanda | x | | x |
| 72. Sao Tome & Principe | | | |
| 73. Saudi Arabia | | | |
| 74. Senegal | x | | x |
| 75. Seychelles | | | x |
| 76. Sierra Leone | x | | x |
| 77. Singapore | x | | x |
| 78. Somalia | * | | x |
| 79. Sri Lanka | x | | x |
| 80. St. Christopher | | | |
| 81. St. Lucia | | | |
| 82. Sudan | x | | x |
| 83. Suriname | | x | x |
| 84. Swaziland | x | | x |
| 85. Syria | x | | x |
| 86. Tanzania | x | | x |
| 87. Togo | | | x |
| 88. Trinidad & Tobago | x | x | |
| 89. Tunisia | x | | |
| 90. Uganda | x | | x |
| 91. United Arab Emir. | | | |
| 92. Vanuatu | | | |
| 93. Vietnam | | | x |
| 94. Yemen Arab Repub. | * | | |
| 95. Yemen, P.D.R. of | x | | x |
| 96. Yugoslavia | x | | x |
| 97. Zaire | x | | x |
| 98. Zambia | x | | |
| 99. Zimbabwe | | | |

x signed and ratified.
* signed only.

# Appendix E

## Treaty Banning Nuclear Weapon Tests in the Atmosphere, in Outer Space and Under Water*

*Signed at Moscow, August 5, 1963*
*Ratification advised by U.S. Senate September 24, 1963*
*Ratified by U.S. President October 7, 1963*
*U.S. ratification deposited at Washington, London, and Moscow October 10, 1963*
*Proclaimed by U.S. President October 10, 1963*
*Entered into force October 10, 1963*

The Government of the United States of America, the United Kingdom of Great Britain and Northern Ireland, and the Union of Soviet Socialist Republics, hereinafter referred to as the "Original Parties,"

Proclaiming as their principal aim the speediest possible achievement of an agreement on general and complete disarmament under strict international control in accordance with the objectives of the United Nations which would put an end to the armaments race and eliminate the incentive to the production and testing or all kinds of weapons, including nuclear weapons.

---

*Source: U.S. Arms Control and Disarmament Agency. Arms Control and Disarmament Agreements: Texts and Histories of Negotiations, 1982 edition. Washington, U.S. Govt. Print. Off., 1982, p. 41-43.

Seeking to achieve the discontinuance of all test explosions of nuclear weapons for all time, determined to continue negotiations to this end, and desiring to put an end to the contamination of man's environment by radioactive substances,

Have agreed as follows:

## Article I

1. Each of the Parties to this Treaty undertakes to prohibit, to prevent, and not to carry out any nuclear weapons test explosion, or any other nuclear explosion, at any place under its jurisdiction or control:

(a) in the atmosphere; beyond its limits, including outer space; or under water, including territorial waters or high seas; or

(b) in any other environment if such explosion causes radioactive debris to be present outside the territorial limits of the State under whose jurisdiction or control such explosion is conducted. It is understood in this connection that the provisions of the subparagraph are without prejudice to the conclusion of a treaty resulting in the permanent banning of all nuclear test explosions, including all such explosions underground, the conclusion of which, as the Parties have stated in the Preamble to this Treaty, they seek to achieve.

2. Each of the Parties to this treaty undertakes furthermore to refrain from causing, encouraging, or in any way participating in, the carrying out of any nuclear weapon test explosion, or any other nuclear explosion, anywhere which would take place in any of the environments described, or have the effect referred to, in paragraph 1 of this Article.

## Article II

1. Any Party may propose amendments to this Treaty. The text of any proposed amendment shall be submitted to the Depositary Governments which shall circulate it to all Parties to this Treaty. Thereafter, if requested to do so by one-third or more of the Parties, the Depositary Governments shall convene a conference, to which they shall invite all the Parties, to consider such amendment.

2. Any amendments to this Treaty must be approved by a majority of the votes of all the Parties to this Treaty, including the votes of all of the Original Parties. The amendment shall enter into force for all Parties upon the deposit of instruments of ratification by a majority of all the Parties, including the instruments of ratification of all of the Original Parties.

## Article III

1. This Treaty shall be open to all States for signature. Any State which does not sign this Treaty before its entry into force in accordance with paragraph 3 of this Article may accede to it at any time.

2. This Treaty shall be subject to ratification by signatory States, instruments of ratification and instruments of accession shall be deposited with the Governments of the Original Parties--The United States of America, the United Kingdom of Great Britain and Northern Ireland, and the Union of Soviet Socialist Republics--which are hereby designated the Depositary Governments.

3. This Treaty shall enter into force after its ratification by all the Original Parties and the deposit of their instruments of ratification.

4. For States whose instruments of ratification or accession are deposited subsequent to the entry into force of this Treaty, it shall enter into force ont he date of the deposit of their instruments of ratification or accession.

5. The Depositary Governments shall promptly inform all signatory and acceding States of the date of each signature, the

date of deposit of each instrument of ratification of and accession to this Treaty, the date of its entry into force, and the date of receipt of any requests for conference or other notices.

6. This Treaty shall be registered by the Depositary Governments pursuant to Article 102 of the Charter of the United Nations.

## Article IV

This Treaty shall be of unlimited duration.

Each Party shall in exercising its national sovereignty have the right to withdraw from the Treaty if its decides that extraordinary events, related to the subject matter of this Treaty, have jeopardized the supreme interests of its country. It shall give notice of such withdrawal to all other Parties to the Treaty three months in advance.

## Article V

This Treaty, of which the English and Russian texts are equally authentic, shall be deposited in the archives of the Depositary Governments. Duly certified copies of this Treaty shall be transmitted by the Depositary Governments to the Governments of the signatory and acceding States.

**IN WITNESS WHEREOF** the undersigned, duly authorized, have signed this Treaty.

**DONE** in triplicate at the city of Moscow the fifth day of August, one thousand nine hundred and sixty-three.

| For the Government of the United States of America | For the Government of the United Kingdom of Great Britain and Northern Ireland | For the Government of the Union of Soviet Socialist Republics |
|---|---|---|
| **DEAN RUSK** | **HOME** | **A. GROMYKO** |

# Appendix F

## Treaty Between the United States of America and the Union of Soviet Socialist Republics on the Limitation of Underground Nuclear Weapon Tests*

*Signed at Moscow July 3, 1974*

The United States of America and the Union of Soviet Socialist Republics, hereinafter referred to as the Parties,

Declaring their intention to achieve at the earliest possible date the cessation of the nuclear arms race and to take effective measures toward reductions in strategic arms, nuclear disarmament, and general and complete disarmament under strict and effective international control,

Recalling the determination expressed by the Parties to the 1963 Treaty Banning Nuclear Weapons Tests in the

---

*Source: U.S. Arms Control and Disarmament Agency. *Arms Control and Disarmament Agreements: Texts and Histories of Negotiations*, 1982 edition. Washington, U.S. Govt. Print. Off., 1982, p. 167-170. This treaty was accompanied by a protocol dated July 3, 1974, that dealt with verification. This protocol was replaced by a protocol signed June 1, 1990, which is reprinted in U.S. Senate. *Protocols to the Threshold Test Ban and Peaceful Nuclear Explosion Treaties with the Union of Soviet Socialist Republics*. Treaty Doc. 101-19. 101st Congress, 2d Session. Washington, U.S. Govt. Print. Off., 1990, p. 7-113.

Atmosphere, in Outer Space and Under Water in its Preamble to seek to achieve the discontinuance of all test explosion of nuclear weapons for all time, and to continue negotiations to this end,

Noting that the adoption of measures for the further limitation of underground nuclear weapon tests would contribute to the achievement of these objectives and would meet the interests of strengthening peace and the further relaxation of international tension,

Reaffirming their adherence to the objective and principles of the Treaty Banning Nuclear Weapons Tests in the Atmosphere, in Outer Space and Under Water and of the Treaty on the Non-Proliferation of Nuclear Weapons,

Have agreed as follows:

## Article I

1. Each Party undertake to prohibit, to prevent, and not to carry our any underground nuclear weapons test having a yield exceeding 150 kilotons at any place under its jurisdiction or control, beginning March 31, 1976.

2. Each Party shall limit the number of its underground nuclear weapon tests to a minimum.

3. The Parties shall continue their negotiations with a view toward achieving a solution to the problem of the cessation of all underground nuclear weapon tests.

## Article II

1. For the purpose of providing assurance of compliance with the provisions of this Treaty, each Party shall use national technical means of verification at its disposal in a manner consistent with the generally recognized principles of international law.

2. Each Party undertakes not to interfere with the national technical means of verification of the other Party operating in accordance with paragraph 1 of this Article.

3. To promote the objectives and implementation of the provisions of this Treaty the Parties shall, as necessary, consult with each other, make inquiries and furnish information in response to such inquiries.

## Article III

The provisions of this Treaty do not extend to underground nuclear explosions carried out by the Parties for peaceful purposes. Underground nuclear explosions for peaceful purposes shall be governed by an agreement which is to be negotiated and concluded by the Parties at the earliest possible time.

## Article IV

This Treaty shall be subject to ratification in accordance with the constitutional procedures of each Party. This Treaty shall enter into force on the day of the exchange of instruments of ratification.

## Article V

1. This Treaty shall remain in force for a period of five years. Unless replaced earlier by an agreement in implementation of the objectives specified in paragraph 3 of Article I of this Treaty, it shall be extended for successive five-year periods unless either Party notifies the other of its termination no later than six months prior to the expiration of the Treaty. Before the expiration of this period the Parties may, as necessary, hold consultations to consider the situation relevant to the substance of this Treaty and to introduce possible amendments to the text of the Treaty.

2. Each Party shall, in exercising its national sovereignty, have the right to withdraw from this Treaty if it decides that extraordinary events related to the subject matter of this Treaty have jeopardized its supreme interests. It shall give notice of its decision to the other Party six months prior to withdrawal

from this Treaty. Such notice shall include a statement of the extraordinary events the notifying Party regards as having jeopardized its supreme interests.

3. This Treaty shall be registered pursuant to Article 102 of the Charter of the United Nations.

**DONE** at Moscow on July 3, 1974, in duplicate, in the English and Russian languages, both texts being equally authentic.

For the United States of America:

**RICHARD NIXON,**

*The President of the United States of America*

For the Union of Soviet Socialist Republics:

**L. BREZHNEV,**

*General Secretary of the Central Committee of the CPSU*

# Appendix G

## Treaty Between the United States of America and the Union of Soviet Socialist Republics on Underground Nuclear Explosions for Peaceful Purposes*

*Signed at Moscow May 28, 1976*

The United States of America and the Union of Soviet Socialist Republics, hereinafter referred to as the Parties,

Proceeding from a desire to implement Article III of the Treaty between the United States of America and the Union of Soviet Socialist Republics on the Limitation of Underground Nuclear Weapon Tests, which calls for the earliest possible conclusion of an agreement on underground nuclear explosions for peaceful purposes,

---

*Source: U.S. Arms Control and Disarmament Agency. *Arms Control and Disarmament Agreements: Texts and Histories of Negotiations*, 1982 edition. Washington, U.S. Govt. Print. Off., 1982, p. 173-189. This treaty was accompanied by a protocol dated May 28, 19976, that dealt with verification. This protocol was replaced by a protocol signed June 1, 1990, which is reprinted in U.S. Senate. *Protocols to the Threshold Test Ban and Peaceful Nuclear Explosion Treaties with the Union of Soviet Socialist Republics*. Treaty Doc. 101-19. 101st Congress, 2d Session. Washington, U.S. Govt. Print. Off., 1990, p. 114-155.

Reaffirming their adherence to the objectives and principles of the Treaty Banning Nuclear Weapons Tests in the Atmosphere, in Outer Space and Under Water, the Treaty on Non-Proliferation of Nuclear Weapons, and their determination to observe strictly the provisions of these international agreements,

Desiring to assure that underground nuclear explosions for peaceful purposes shall not be used for purposes related to nuclear weapons,

Desiring that utilization of nuclear energy be directed only toward peaceful purposes,

Desiring to develop appropriately cooperation in the field of underground nuclear explosions for peaceful purposes,

Have agreed as follows:

## Article I

1. The Parties enter into this Treaty to satisfy the obligations in Article III of the Treaty on the Limitation of Underground Nuclear Weapon Tests,, and assume additional obligations in accordance with the provisions of this treaty.

2. This Treaty shall govern all underground nuclear explosions for peaceful purposes conducted by the Parties after March 31, 1976.

## Article II

For the purposes of this Treaty:

(a) "explosion" means any individual or group underground explosion for peaceful purposes;

(b) "explosive" means any device, mechanism or system for producing an individual explosion;

(c)"group explosion" means two or more individual explosions for which the time interval between successive individual explosions does not exceed five seconds and for which the emplacement points of all explosives can be interconnected by straight line segments, each of which joins two emplacement points and each of which does not exceed 40 kilometers.

## Article III

1. Each Party, subject to the obligations assumed under this Treaty and other International agreements, reserves the right to:

(a) carry out explosions at any place under its jurisdiction or control outside the geographical boundaries of test sites specified under the provisions of the Treaty on the Limitation of Underground Nuclear Weapon Tests; and

(b) carry out, participate or assist in carrying out explosions in the territory of another State at the request of such other State.

2. Each Party undertakes to prohibit, to prevent and not to carry out at any place under its jurisdiction or control, and further undertakes not to carry out, participate or assist in carrying out anywhere:

(a) any individual explosion having a yield exceeding 150 kilotons;

(b) any group explosion;

(1) having an aggregate yield exceeding 150 kilotons except in ways that will permit Identification of each individual explosion and determination of the yield of each individual explosion in the group in accordance with the provisions of Article IV of and the Protocol to this Treaty;

(2) having an aggregate yield exceeding one and one-half megatons;

(c) any explosion which does not carry out a peaceful application;

(d) any explosion except in compliance with the provisions of the Treaty Banning Nuclear Weapon Tests in the Atmosphere, in Outer Space and Under Water, the Treaty on the Non-Proliferation of Nuclear Weapons, and other international agreements entered into by that Party.

3. The question of carrying out any individual explosion having a yield exceeding the yield specified in paragraph 2(a) of this article will be considered by the Parties at an appropriate time to be agreed.

## Article IV

1. For the purpose of providing assurance of compliance with the provisions of this Treaty, each Party shall:
(a) use national technical means of verification at its disposal in a manner consistent with generally recognized principles of international law; and
(b) provide to the other Party information and access to sites of explosions and furnish assistance in accordance with the provisions set forth in the Protocol to this Treaty.

2. Each Party undertakes not to interfere with the national technical means of verification of the other Party operating in accordance with paragraph 1(a) of this article, or with the implementation of the provisions of paragraph 1(b) of this article.

## Article V

1. To promote the objectives and implementation of the provisions of this Treaty, the Parties shall establish promptly a Joint Consultative Commission within the framework of which they will:

(a) consult with each other, make inquiries and furnish information in response to such inquiries, to assure confidence in compliance with the obligations assumed;

(b) consider questions concerning compliance with the obligations assumed and related situations which may be considered ambiguous;

(c) consider questions involving unintended interference with the means for assuring compliance with the provisions of this Treaty;

(d) consider changes in technology or other new circumstances which have a bearing on the provisions of this Treaty; and

(e) consider possible amendments to provisions governing underground nuclear explosions for peaceful purposes.

2. The Parties through consultation shall establish, and may amend as appropriate, Regulations for the Joint Consultative Commission governing procedures, composition and other relevant matters.

## Article VI

1. The Parties will develop cooperation on the basis of mutual benefit, equality, and reciprocity in various areas related to carrying out underground nuclear explosions for peaceful purposes.

2. The Joint Consultative Commission will facilitate this cooperation by considering specific areas and forms of cooperation which shall be determined by agreement between the Parties in accordance with their constitutional procedures.

3. The Parties will appropriately inform the International Atomic Energy Agency of results of their cooperation in the field of underground nuclear explosions for peaceful purposes.

## Article VII

1. Each Party shall continue to promote the development of the international agreement or agreements and procedures provided for in Article V of the Treaty on the NonProliferation of Nuclear Weapons, and shall provide appropriate assistance to the International Atomic Energy Agency in this regard.

2. Each Party undertakes not to carry out, participate or assist in the carrying out of any explosion in the territory of another State unless that State agrees to the implementation in its territory of the international observation and procedures contemplated by Article V of the Treaty on the Non-Proliferation of Nuclear Weapons and the provisions of Article IV of and the Protocol to this Treaty, including the provision by that State of the assistance necessary for such implementation and of the privileges and immunities specified in the Protocol.

## Article VIII

1. This Treaty shall remain in force for a period of five years, and it shall be extended for successive five-year periods unless either Party notifies the other of its termination no later than six months prior to its expiration. Before the expiration of this period the Parties may, as necessary, hold consultations to consider the situation relevant to the substance of this Treaty. However, under no circumstances shall either Party be entitled to terminate this Treaty while the Treaty on the Limitation of Underground Nuclear Weapon Tests remains in force.

2. Termination of the Treaty on the Limitation of Underground Nuclear Weapon Tests shall entitle either Party to withdraw from this Treaty at any time.

3. Each Party may propose amendments to this Treaty. amendments shall enter into force on the day of the exchange of instruments of ratification of such amendments.

## Article IX

1. This Treaty including the Protocol which forms an integral part hereof, shall be subject to ratification in accordance with the constitutional procedures of each Party. This Treaty shall enter into force on the day of the exchange of instruments of ratification which exchange shall take place simultaneously with the exchange of instruments of ratification of the Treaty on the Limitation of Underground Nuclear Weapon Tests.

2. This Treaty shall be registered pursuant to Article 102 of the Charter of the United Nations.

**DONE** at Washington and Moscow, on May 28, 1976, in duplicate, in the English and Russian languages, both texts being equally authentic.

For the United States of America:

**GERALD R. FORD,**

*The President of the United States of America.*

For the Union of Soviet Socialist Republics:

**L. BREZHNEV,**

*General Secretary of the Central Committee of the CPSU*

# Appendix H

## Letter of Introduction

House of Representatives
Committee on Foreign Affairs
*Washington, D.C., June 16, 1989*

As part of the Foreign Affairs Committee's ongoing interest in arms control, Congressman William S. Broomfield, the Vice-Chairman of the committee, joined our friend and colleague on the Committee on Armed Services, Congresswoman Beverly Byron, and requested that the Congressional Research Service prepare the following report on nuclear testing. This report offers a detailed analysis of this important issue, an issue which is of concern to all of us.

The analysis and finding contained in this report are those of the Congressional Research Service and, as such, they do not necessarily reflect the views of the Committee on Foreign Affairs or its members.

Dante B. Fascell, *Chairman.*

# Appendix I

## Letter of Submittal

Congressional Research Service
The Library of Congress
*Washington, D.C., March 31, 1989*

Hon. Beverly Byron,
*Chairwoman, Subcommittee on Military Personnel and Compensation, Committee on Armed Services, House of Representatives, Washington, D.C.*

Hon. William S. Broomfield,
*Ranking Minority Member, Committee on Foreign Affairs, House of Representatives, Washington, D.C.*

Dear Representatives Byron and Broomfield: I am pleased to submit to you the enclosed study on nuclear testing issues that you requested.

You asked that the study consider "*all* aspects of the nuclear testing issue in order to create an objective, factual and analytical basis for congressional consideration of various nuclear test ban and testing moratoria initiatives." To this end, the study is structured as follows. Chapter 1 introduces the study and summarizes our findings. Chapter 2 presents the history of the test ban debate. Chapter 3 is a description of the technology of nuclear warheads, and United States and Soviet nuclear stockpile, and the U.S. and Soviet testing programs. Chapter 4 details how various test limitations would affect the development of new nuclear warheads and thereby U.S. security. Chapter 5 discusses stockpile confidence under various test limitations and the balance between effects on the United States and the Soviet Union. Chapter 6 considers how alternate test bans would impinge on U.S. ability to determine how nuclear detonations affect U.S. military systems. Chapter 7 describes issues associated with verification of compliance with various test limits. Chapter 8 assesses how various testing limitations might affect nuclear powers other than the United States and Soviet Union, nonnuclear allies, and the proliferation of nuclear weapons.

The Department of Energy has reviewed the study and has certified that it does not contain classified material.

Sincerely,

Joseph Ross, *Director.*

# Appendix J

## Letter of Transmittal

June 8, 1989.

Hon. Dante B. Fascell,
*Chairman, Committee on Foreign Affairs, House of Representatives, Washington, D.C.*

Dear Dante: Attached is a study prepared by the Congressional Research Service of the Library of Congress entitled, *Nuclear Weapons and Security: The Effects of Alternative Test Ban Treaties.*

I believe this is the most thorough and comprehensive unclassified analysis of the Nuclear Test Ban issue prepared to date. Consistent with CRS's mandate, this study is objective and deals in a detailed manner with this very important issue.

I jointly requested this study with Congresswoman Byron of the Armed Services Committee, but because I initiated this study I would like to see the Committee on Foreign Affairs publish this report.

Dante, I would appreciate your approval for printing this report as a Committee on Foreign Affairs committee print. Thank you very much for your attention to this matter.

Sincerely,

William S. Broomfield,
*Ranking Minority Member.*